A SURGEON'S ODYSSEY

THROUGH WORLD WAR I AND POST-ARMISTICE MISSIONS

A SURGEON'S ODYSSEY
THROUGH WORLD WAR I AND POST-ARMISTICE MISSIONS

A COMPILATION OF **ESSAYS** AND **LETTERS**
BY DR. GEORGE DE TARNOWSKY

EDITED BY
WILLIAM JEFFREY HUDSON JR. AND NICOLE HUDSON

First print and eBook edition October 2024

ISBN Paperback: 978-1-964756-00-4
ISBN Hardcover: 978-1-964756-01-1
ISBN eBook: 978-1-964756-02-8

Published by Simply Hudson Publishing

Table of Contents

Introduction

by
William J. Hudson, Jr.

After my father passed away, my sister came across an old envelope in a dresser drawer. The envelope contained numerous typewritten pages regarding World War I. While she had no interest in them, she contacted me and, seeing as I was not familiar with the contents, I asked her to send the envelope my way.

Upon receiving the package, I discovered that the envelope contained a series of essays and short stories based on the firsthand experiences of my grandfather. Crafting his words with a blend of artistic prowess inherited from both his Russian-Polish father and American mother, Dr. George de Tarnowsky had written these accounts, providing a unique perspective on the events of World War I and the Post-Armistice. Graduated from medical school at Northwestern University, the French-born immigrant was a prominent surgeon in the City of Chicago.

We are aware that Dr. de Tarnowsky had received orders from the Surgeon General, during the summer following the United States' entry into the war in April 1917, to survey the various approaches employed in caring for the wounded. This endeavor led to the publication of a manual, titled *Military Surgery of the Zone of the Advance*, on establishing aid stations and surgical centers. Initially, his mission as a Medical Reserves Officer encountered challenges due to a lack of support and access, hindering his work. Furthermore, it is understood that Dr. de Tarnowsky diligently composed these writings on a typewriter, capturing his thoughts on double-quick, amidst the rapid-fire changing landscape of emergency war surgery and medicine during the tumultuous period of Western Front trench warfare. Subsequently, during the post-Armistice era, he wrote reflectively and philosophically while traveling by train or staying in temporary lodgings. These essays stemmed from ethnographic and phenomenological observations, distinguishing them from conventional laboratory work and literature research.

Throughout WWI, Dr. de Tarnowsky served as a dedicated war surgeon, conducting extensive research and collaborating closely with both the American and International Red Cross. He diligently took fieldnotes and ethnographic records of his interactions with a diverse range of nationalities, including the French, British, Germans, Belgians, Serbians, Romanians, Italians, Croatians, and fellow Americans. It is remarkable how he swiftly typed on a typewriter, capturing intricate details and providing vivid insights. His recorded words guide us through visceral hardships of the unpredictable trench warfare and the post-Armistice landscape.

Fearlessly, he ventured to the frontlines, braving the dangers, to gather invaluable data on medical procedures and acquire knowledge of war surgical methods from the French and British. As we attentively listen to his voice, we are transported to the harrowing realities of the Western Front trenches. The cadence of Dr. de Tarnowsky's words resonates, offering profound insights into a cultivated order that reveals the interlocking patterns of purpose, genteel breeding, customs/culture, family, and philosophy, both within the warfare landscape and his interactions with others on the battlefield and post-Armistice settings.

Dr. de Tarnowsky's writing style was unique. He combined observational writings such as journal entries and detailed accounts with reports, orders, extracts, letters, and witness testimonies. His "George Skitarnow essays" interweave with the observational writings, forming an engaging ethnographic storytelling experience that educates us about diverse cultures, including the tea-drinking British, the "C'est la Guerre" French, the Zacouska-loving Romanians, among others. Additionally, his satirical fiction stories provide a much-needed comedic relief amidst the horrors of the distinctive environment forged by the trench warfare of World War I.

With careful editorial deliberation, we preserved Dr. de Tarnowsky's distinctive voice, embracing his multilingual command of the language and communication. The flaws of hyphens and spacing in words, misspellings, and punctuation misplacements only deepen the charm of his writings. These imperfections reflect the limitations of typewriters and the reliance on thorough study and reading knowledge for writing and editing, as opposed to today's automatic editorial assistance. Such factors inspire us and highlight Dr. George de Tarnowsky's intelligence, which emanates from his military and medical studies and research, as well as his profound commitment to altruism and diplomacy for the well-being of others.

After re-reading my grandfather's essays, I am again impressed by his language, literary, and observational skills which have provided deeper insight into the man I once knew. While I wish I had the opportunity to know him better during his lifetime, I am grateful for the times that we did have together and the legacy he has left behind.

In closing, I would like to thank Dr. Sarah Elizabeth Moreman, Nicole Hudson, and William Jeffery Hudson de Tarnowsky for their support in bringing this project to fruition.

WWI and de Tarnowsky's Odyssey

April 1917 - U.S. declared war on Germany

June 1917 - First Division U.S. troops landed on France

July 1917 - George de Tarnowsky undertook the assignment

September 1917 - pursued task to write War Manual by crossing the Atlantic on *S.S. St. Louis*

October 3, 1917 - arrived in Paris

October 17, 1917 - Base Hospital No 6 A.E.F., Bordeaux

October 23, 1917 - proceeded to Chateau d'Annel Hospital, Ambulance des Alliès

November 12, 1917 - authorizations to visit French Sanitary formations in the trenches

December 24, 1917 - Cassel, Flanders fields

December 28, 1917 - Journal studying French Bataillon

January 1, 1918 - wartime New Year's dinner

January 29, 1918 - report on Sanitary Service of French zone of the Advance showed that he earned the letters M.C.

February 26, 1918 - "Somewhere in France"

March 10, 1918 - final tour investigation completed

March 13-20, 1918 (approximately) - a week's leave to visit sister in Nice

March 21, 1918 - Kaiser's Battle

March 22, 1918 - returned from Paris to Chateau d'Annel

March 25, 1918 - Directrice ordered evacuation of the hospital

March 26-27, 1918 - pulled all-nighter to perform 26 consecutive operations

March 27, 1918 - Germans suspected trap after seeing the 15-mile gap between British and French armies

March 30, 1918 - "Somewhere in France" published in Journal of American Medical Association

April 3, 1918 - organized three operating teams with ten-hour shifts

May 26, 1918 - the Chemin-des-Dames surprise attack by Germans

May 27, 1918 - end of Ambulance des Alliès

after May 27, 1918 - gained command of another Red Cross Military Hospital - No 5

June 1918 - War Manual No 7 out for distribution, nine months after arrival in France

July 18, 1918 - turning point of the war

August 8, 1918 - Black Day

November 11, 1918 - "Cease Firing" order / Armistice Day

January 1919 - Serbian Red Cross medal

January 24 - February 11, 1919 - Serbia

May 14-24, 1919 - Left Bologna for Marseilles

June 26 - July 15, 1919 - Left Marseilles for Bucharest

July 16-27, 1919 - Bucharest

July 27 - August 1, 1919 - Belgrade

August 9, 1919 - back in Paris

April 1920 - French Decoration

April 1920 - Roumanian Decoration

About 1920 - Serbian Decoration

June 23, 1922 - Distinguished Service Medal

1925 - Medical Book of Merit

May 31, 1930 - The Six Hundredth Jubilee

Armistice Day 1938 - Prevention and preparation speech at the University of Illinois

Glossary of Terms

C'est la guerre	French expression meaning "It's the war!"
A.E.F.	American Expeditionary Forces
Abri des permissionaires	Place of rest for soldiers on furlough
Adjutant General	The chief administrative officer in the United States Army; adjutant of a unit having a general staff
Albergo	Italian for hotel or inn
Amex, Sam Ex, Sammies	Favorite nickname overseas for American Expeditionary Forces
Anzac	Australian and New Zealand divisions of the British army
Auctioneering	Auctioning of goods such as cigarettes and beer
Auto-Chir	Automobile Chirurgical or specialized surgical hospitals rapidly moved on trucks from sector to sector
A.W.O.L.	Absent without leave
Banat	Roumanian region or dialect
Batman	Orderly
Billets	A place, usually a civilian's house or other non-military facility, where soldiers are lodged temporarily
Blighty	England on leave
Boches	Derogatory term for German soldiers
Caisson	Chest or wagon for holding or conveying ammunition; casket of the fallen military member drawn by four harnessed horses
Camion	Truck or lorry used to transport goods by road, for carrying freight or military supplies, equipment, and troops
Caressa	The characteristic vehicle of Rome, antiquated in appearance with small wheels in front and a huge umbrella to be hoisted over the driver's head in bad weather, and with very small horses
Chillblains	A condition that causes inflamed swollen patches and blistering on the hands and feet; usually caused by exposure to damp air that's cold but not freezing

Cogitating	Think deeply about something, meditate, or reflect
Cooperative	Canteen or commissary to purchase goods and groceries
Crapouillot	"Crapaud" in French, "little toad" used in the trenches, the standard French medium trench mortar, a French magazine
C.S.S.	Casualty cleaning stations (C.S.S.) or evacuation hospitals that move from place to place, making communications impossible until the wounds are healed
Cuistot	French term for chef or cook
Demi-tasse	Chickory mixed with coffee, virgin of roasted bean, roast barley
Directrice	Madame la Chatelaine (wealthy New York expatriate); a woman who directs or controls as head or director of an institution, in this case the Chateau d'Annel Ambulance des Alliès
Double-quick	Very quickly
Duckboards/duck-boards	The bottom of trenches was overlaid with sections of wooden crossbars, two yards long, built ladder fashion with cross-sticks two feet wide and two inches apart which prevented the soldiers from slipping and floundering about in the mud during the rainy season. The English called these sections duck-boards, the French and Belgians "caillebottis"
En pensione	Fixed rate for board and lodging (meals provided); pensione means an Italian boarding house
Enfilading fire	Direct a volley of gunfire along the length of a target
Épris	Enamoured
Fag	Smoke, cigarette
Feldgrau	Grayish-green color of military uniforms of German armed forces for camouflage as a military tactic
Feldwebel	Sergeant, a noncommissioned officer
Flanders	Muddy flats of western Belgium
Fritz	German soldiers taking positions in the trenches, derived from the development of Fritz Haber of Germany's poison-gas program synthesis of ammonia from hydrogen and nitrogen
Hopital origine ètapes (H.O.E.)	French evacuation hospital

Hopitaux Bènèvoles	Amateur hospitals that sprung up in 1914 as a gesture of goodwill towards France
Hillock	Small hill or mound
Hinterland	German for land left behind
Il Boore	Local term or regional name for a strong, gusty wind that originates from the mountains
Imprecations	Spoken curse
Irridenta	A region that is related ethnically or historically to one country but is controlled politically by another
Jaeger underwear	Pure, sanitary wool underthings to help ward off cold and heat, economical form of dress
Je suis jereste	French for "here I am, here I stay"
Kommandatur	Local enemy (German) being the Kommandant that is a commandant/officer who has the command of a place or of a body of troops, thus the Kommandatur is the headquarters or office of the Kommandant
Lighter	A large, flat-bottomed barge used in loading or unloading ships.
Limogè	French term to designate the fate of officers whose active services were no longer needed
Litter-drill	Complicated and now obsolete series of maneuvers carried out by stretcher-bearers which only ended when certain number of stretchers were neatly stacked up at regular intervals
Marraines	War godmothers
Mitrailleuse cages	Carabine multiple, an early Franco-Belgian machine gun, volley gun, human war engine - man and machine gun almost indistinguishable
Musette	Canvas bag with a leather bottom as light alternative to a full pack used to carry small equipment or personal items during WWI
Niçois peasants	Soldiers of Nice, a French city on the Mediterranean coast
Oubliettes	A secret dungeon with access only through a trapdoor in its ceiling
Patois	Jargon, vernacular, or informal speech used by a particular group of a region

Pèpères	Fathers, men in their 50s inducted into service because France almost "bled white" by Germans, usually served as scrubmen or stretcher-bearers due to no previous hospital training
Permissionaire	Soldier on furlough
Pill-box	Type of blockhouse, or concrete dug-in guard post, as a miniature fort with loopholes through which defenders can fire weapons; a trench firing method developed by Germans to strengthen trench lines
Pinard	French slang/Poilu term for wine (specifically red wine) during WWI
Poilus	"Hairy ones" - French comrades or soldiers who had no time for shaving during the agonizing weeks of August 1914
Primus stove	The first soot-free, pressurized-burner kerosene (paraffin) stove, developed in 1892 by Frans Wilhelm Lindqvist and J.V. Svenson in a small blacksmith workshop in central Stockholm; as it was the first, they called it Primus (meaning first)
Puttees	8 feet long khaki-colored cloth such as canvas or wool as covering to wrap around the lower part of the leg from ankle to knee, leg wraps to be worn between boots and breeches; provide support and protection against harsh weather conditions, to prevent debris and water from entering boots
Rabbit warrens	Dug-outs
Rainbow Division	The 42nd Infantry Division 42ID of the United States Army National Guard; the fourth American unit to land in France; formed by taking National Guard units from 26 states and combining them into a division that stretched across the country, "like a rainbow," in the words of the division chief of staff, Army Col. Douglas MacArthur
Rata	Classic, historic French army ration; concoction of demi-soup and demi-stew with mysterious seasonings and wine sauce
Requisition	Official order laying claim to the use of property or materials
Saccharin	French WWI substitute for sugar
Salient	Bulge in the line that gives the defenders the advantage

	of being able to fire on those in the salient from three sides; a battlefield feature that projects into enemy territory
Sally	Derived from Latin term salire (to jump) or salle sortie, a military maneuver typically during a siege, a military action in which besieged troops burst forth from their position; soldiers on defensive, having retreated to a foxhole or fort, make an abrupt offensive attack on their opponents; sudden rushing forth, a quick raid
Sam Browne belt	Leather belt with supporting strap over the right shoulder; named after Sir Samuel J. Browne (1824-1901), the British Indian army general who invented this to be able to carry a sword
Sava River	A river in central and southeast Europe, the longest tributary of the Danube, flowing through Slovenia, Croatia, alongside Bosnia and Herzegovina, and finally Serbia
Shock troops	A group of soldiers trained specially for carrying out a sudden assault
Sortie	Term originated in siege warfare: a deployment or dispatch of one military unit, be it an aircraft, ship, or troops, from a strongpoint; can be used interchangeably with another term, "sally"
Strafing	Attack repeatedly with bombs or machine gunfire from low-flying aircraft
Terpsichorean	Relating to dancing
Teutons	Gallic words "teuta" means tribe and "ona" means water = Germans/Germanic tribe from Jutland
Tommies, Tommy	Slang for British soldier(s)
Traceries	Architectural detail in ornamental stone openwork, typically in the upper part of a Gothic window
Triage	Accurate sorting out process
Tzigane	Rhapsodic composition by the French composer Maurice Ravel for violin and piano
Umbria	Central Italy, revered as the country's green heart and known for its medieval hill towns, local cuisine famous for truffles, chocolate, and wine, and lush landscapes of dense forests and olive groves, bordering Tuscany, Lazio, and Le Marche

Unlimber	Detach a gun from its limber so that it can be used
Vicissitude	A change of circumstances, typically unwelcome or unpleasant
Zacouska	Traditional Roumanian roasted vegetable spread or vegetable caviar typically consisting of roasted eggplant, roasted peppers, tomatoes, and onions
Zeppelins	French nickname "Gotha" - rigid-framed German airship that retains its shape with or without gas; once used as people carriers, turned into weapons of war
Zuica	Traditional Romanian spirit containing ~24-86% alcohol by volume prepared from plums; national liquor distilled from prunes

Preface

Twenty Years Later

"Armistice Day" 1938 – Prevention and Preparedness

Friday, November 11, 1938

As a famous French surgeon once told your speaker, with the cool logic which is so typical of that race: "In the zone of combat one does what one can, where one can, when one can, how one can!"

Speech at University of Illinois School of Medicine

On November 11th, 1918, the order "Cease Firing" resounded from Flanders fields to the Swiss frontier. Millions of men of all arms and branches of the service and countless thousands of devoted young women who had been braving cold, hunger, fatigue, or death began to wonder how they were going to readjust themselves to normal, humdrum every-day civilian life! Not that any of them regretted the cessation of hostilities; on the contrary, one could almost hear the sigh of relief which replaced the roar of cannon, whistling of bullets, and zooming of aeroplanes. From the front lines to the rear, from Europe to the United States, joy manifested itself in an ever increasing crescendo. We, overseas, were even told that a premature Armistice celebration had so depleted the supply of telephone books and ticker-tape in the United States, that when the real day arrived, it became a hard task to fill the city's streets with scrap paper! In the zone of the advance as it was then called, or combat zone as it is now designated, we were dazed by the sudden deathless silence around us, a silence soon broken by shouting, whistling, cat-calling, or rattling of tin cups or pans as our wounded boys manifested as best they could the beginning of a new era. No more camouflage, no darkened wards, dressing rooms, or operating pavillions; all secrecy removed, the sky clear of enemy planes, peace, peace all around us. Finie la guerre!

The war is ended! For us of the Medical and nursing corps, there was plenty more work to be done before the last wounded buddy had been patched up and shipped to base hospitals via American hospital trains. The aftermath of the holocaust is still with us, our Veteran's hospitals grim with reminders that war wounds often leave permanent scars which only death can remove.

We are gathered to-day to pay just tribute to the millions of young Americans who responded to the call in 1917, who hurriedly and only too often partially trained, made up the personnel of the American Expeditionary Forces whose combined efforts wrote one of the most thrilling pages in the history of our country. We have listened to the roll-call of the men of the U. of Ill who made the supreme sacrifice; we have just bowed our heads and faced East while listening to the mournful yet beautiful bugle notes; let us now briefly review what they, in the living, and we who were fortunate enough to return home more or less sound in body and mind, accomplished overseas. Unprepared as we were, improvising as we went along, having only the bare rudiments of a General Staff, obliged to buy or borrow from our Allies a large, very large percentage of necessary war materiel, we nevertheless accomplished more than was expected of us and certainly were a strong factor in permanently turning the tide of battle. Few if any of you, present here to-day, know that the American Army in France was not expected to take the field in force before the early months of 1919 – yet such was the plan of the Allied War Council. The enemy upset all of these conservative calculations by his sudden onslaught in March 1918 when he sought to, and almost succeeded in, separating the French from the British armies. Lloyd George, Prime Minister of England at that time, could truly exclaim in the House of Commons: "We now have our backs to the wall!" Few of you know that, on March 27th, 1918, there was a 24 km gap (15 miles) between the French and British armies where not a single soldier remained to hold the line. The enemy saw this gap, suspected a trap, and stopped their advance in this sector, thus giving the French time to march up and take over what the British had abandoned! March, April, and May 1918 were anxious months indeed for the Allies. It was during those critical weeks that General Pershing placed the entire American Expeditionary Forces at the disposal of Marshall Foch, wisely stipulating however that our units were not to be broken up and almagamated with Allied units but were to remain in command of our own officers. Only four of our divisions had at that time received the baptism of fire, the First, Second, Twenty-sixth, and Forty-second. The first division was composed entirely of regulars, the second was a mixed unit with one brigade of regulars and our brigade of marines; the 26th was a New-England division composed of National guardsmen, commanded in the staff grades by officers of the regular army, while the 42nd represented a grouping of companies selected for their military aptitude from all of our training camps. It naturally came to be known as the Rainbow Division. These four sterling units were from the beginning, and remained to the end of the war, our shock troops, called upon whenever other units failed in attaining their objectives. No criticism is intended in this statement; subsequent divisions as they arrived in France and were tested, were found to lack, to a greater or lesser degree, experience, cohesion, and esprit-de-corps, absolute requirements for any military organization. It is one thing to march in alignment before visiting dignitaries, to "eyes right" smartly at the word of command or to assume "parade rest" attitude while listening to platitudes – and entirely another matter to be suddenly confronted with situations which cannot possibly be covered in any manual of instructions! Some of my audience may have read the dramatic story of the "Lost Battailon". The incident occurred during the Argonne drive which began on the 25th of September 1918. This is how the authors of the book described the men of the 1st battalion, 308th regiment, 77th division: "The men had no blankets, no raincoats, no reserve rations AND NO EXPERIENCE! They were draft boys from New-York city's lower East side; only 10% of them knew how to work a hand grenade, and as for the rifles "I can't make the bullets go into this thing" said one! But the orders were: "Go ahead, pay no attention to flanks or losses". That this battalion was finally extricated after five days of stubborn resistance is another part of the story; the point I wish to drive home is the lack of training of many of our soldiers, even as late as October 1918. Time after time, during inspection of the wards in the Evacuation hospital I had the honor to command, I would quiz the wounded in order to find out why and how they had been wounded. Time after time, the answer would be: "I was walking through a wheat field. Why weren't you crawling on your belly? Nobody had told me how to do so!" Many of these poor replacements had arrived in France, especially during the last weeks of the war, after a few days in camp, minus the rudiments of military tactics. They sometimes didn't seem to know the difference between the butt-end of the gun and the bayonet! Such conditions must never again be allowed to recur – and they won't recur if the National Guard and Organized Reserves of the country continue with the present plan of periodic training!

Even we, of the Medical corps, had to improvise. The military surgeon at the front is no longer able to choose his own conditions; he has to make the best of those in which his lot is cast. As a famous French surgeon once told your speaker, with the cool logic which is so typical of that race: "In the zone of combat one does what one can, where one can, when one can, how one can!" Perhaps one or two incidents will drive home what I mean by adaptation to circumstances during periods of stress. Throughout our short course of training in camp, particular stress was laid on "litter-drill", a complicated and now obsolete series of maneuvers carried out by stretcher-bearers which only ended when a certain definite number of stretchers were neatly stacked up at regular intervals, with the enlisted men standing at attention while the unit commander saw to it that each stretcher leaned at a certain angle so that these stacks looked exactly like Indian tepees. During the rush of one of our stiffest battles, our hospital trains were sore pressed to keep the Evacuation hospitals cleared for more front-line admissions. Rapid loading, transit, and unloading were imperative if all our wounded were to receive adequate attention. A certain Captain, M.C., fresh from training camp, came marching up to the station platform in excellent military formation, halted his men and then, instead of dividing his squads so that the train of wounded could be rapidly unloaded, he ordered the classical "litter-drill". The commanding officer of the hospital train spied the maneuver, leaned out of a car window, and fairly screamed: "Stop that blankety-blank drill at once and empty my train!" In a certain Evacuation Hospital, at the height of the Chateau-Thierry, Belleau woods battles, when every cot was occupied, twenty-two operating tables were in constant use, night and day, and all hospital corps men were sorely needed, one of the surgeons, a recent arrival of course, wished to operate upon his orderly for a potential hernia and could not at first understand why his operation was sternly countermanded! Wounded, first, last, and all the time; none but acute surgical abdomens among the personnel could be operated upon!

These are two of the rare instances where individuals failed to early adapt themselves to actual conditions. Many, many more tales can be told of quick decisions, inspiring courage, and extraordinary determination on the part of officers and men of the medical and dental corps, as well as on the part of members of the nursing corps. The following are quotations from official records:"78th Division, Thiaucourt sector. Two enlisted men, M.C. were administering first-aid to a wounded soldier, when they were captured by the enemy. While being marched to the German rear by a sentry, they jumped the sentry, disarmed him and brought him and their wounded buddy back to the American lines.

An ambulance driver was under his ambulance fixing defective wires when an enemy shell crashed through the car spilling burning gazoline on him and his breeches. He became very indignant because he was not immediately given a new car and new breeches!"

Let me now recount the heroic conduct and praiseworthy initiative of a Lieutenant, Dental Corps, who, like hundreds of other dental officers in the A.E.F., became surgeons overnight, gave anesthetics, dressed wounds, applied splints, and even performed emergency amputations. I quote: "When the 35th Division Infantry was withdrawn by Corps orders from Excermont sector, nearly all the wounded were carried or sent back. About 225 were collected at Chaudron farm. Lt Hine, Dental Corps, then on duty with the 2nd battalion, 139th Infantry, receiving information that a number of wounded were left at the farm under guard and that a medical officer was needed, proceeded to that place of his own accord, accompanied by his enlisted dental assistant Wm C. Myers. During the afternoon of September 29th, he worked continuously, dressing and evacuating patients, constantly under fire of high explosives and gas shells. Twice during the day, the Germans attacked the dressing station and were driven back from the very walls by a detachment of the 139th Infantry under Captain Randell Wilson. At 10 P.M., September 29th, there still remained 95 litter patients. The guard under Captain Wilson had suffered heavy losses and, in spite of requests, no aid had been received. Lt Hine was informed by Captain Wilson that the main line of battle had fallen back a half-mile behind the aid station and that there were only 25 Infantrymen between the station and the Germans. To secure assistance, Lt Hine voluntarily procured a mule which was in the barn of the farm and, riding bareback with a rope halter, started back through heavy artillery fire and machine-gun fire for help. Lt Hine arranged for assistance by ambulance, litters, and litter-bearers; he then started back to his station always under shell fire. Believing however that all of the wounded could not be removed before daylight and knowing that the Infantry protection was inadequate, Lt Hine again mounted his mule and proceeded under fire to Artillery H.Q. to procure a protective barrage in front of the station. Knowing that the Infantry guard might be out in front where the barrage had been ordered to be placed, Lt Hine ran his mule back through fields and hills to the station, which was at this time a half-mile in front of the front lines, and succeeded in getting the Infantry back just before the barrage started. At daylight, September 30th, there were only 12 men of the guard left and a number of wounded still unevacuated. Lt Hine therefore again made a trip back to procure more litters and litter-bearers, and with Lt Deal, took the men and litters over the hill towards the station under a most violent shell-fire. He continued his work until every soldier as well as five German prisoners had been evacuated and the aid-station entirely cleared. For this piece of extraordinary gallantry, Lt Hine received the Distinguished Service Cross.

Lest the Dental corps might feel that it had been particularly singled out, let me now recount the deed of a Reserve Army surgeon, Lt Bret Vernon Bates, now practicing in Brown's Valley, Minnesota. I quote: "On the morning of September 30th, in order to expedite the evacuation of wounded groups and scattered wounded from the front, Lt Bates, assisted by Lt R.T. Speck, both from Ambulance Co 139, organized detachments of litter-bearers and with them and eight mule-drawn ambulances went out over the open fields, under heavy artillery and machine-gun fire, in plain sight of the enemy, to collect and evacuate the wounded. These detachments and mule ambulances went cooly on with their work in spite of casualties and in spite of the fire, and even continued during an enemy counter-attack. The mule ambulances drove over the fields when the infantry was held closely in shell holes and trenches on account of the fire. Lt Bates was likewise awarded the Distinguished Service Cross.

It would be manifestly unfair to mention heroism among Dentists and Surgeons and omit recounting the courage and self-sacrifice of at least one member of the nursing corps. On May 18th, 1918, the French advanced surgical center of Beaurieux, in Chemin-Des-Dames sector asked for and obtained the "loan" of one of your speaker's best surgical nurses, Miss Colt Williams. On the above date, the Chemin-des-Dames sector was so quiet that it was looked upon as a rest area for tired troops. Twelve days later, in this same sector, the Germans launched the most complete and efficient surprise attack ever seen on the Western front, an advance which was only halted after they had crossed the Marne at Chateau-Thierry, and our Marines had begun that glorious stubborn step-by-step fight through Vaux village, & Belleau woods which threw the enemy across the river and began to crack his morale. At nine o'clock on the morning of May 27th, a French orderly rushed in the ward where Miss Williams was doing surgical dressings, shouting: "The Germans are here!" Hardly had he uttered his warning cry when in strode a German Colonel M.C. followed by several junior surgeons. The entire hospital was captured. During the German occupancy of the hospital and, later on, when the German retreat had begun, Miss Williams was kept at her task, instead of being sent to a prisoner's war camp in Germany. Shifted from one field hospital to another, from dressing station to dressing station, she was compelled to dress wounds out in the open, under heavy fire, badly fed, miserably housed, lacking even elementary conveniences such as any Anglo-Saxon is accustomed to. Shortly before the Armistice, Miss Williams was finally placed on a freight train and taken to Camp Ruhleben, Germany, where she was interned. Dysentery had, by that time, sapped her strength. When she received the coveted Croix de Guerre in the court of honor of the Invalides in Paris, late in November 1918, she was a mere shadow of her former self. A month later she died, one of the many unsung heroines of the war! On the day of her military funeral, her fiancé, a French major M.C. committed suicide!

We of the Medical, Dental, and Nursing corps have reason to be proud of our efforts during the World War. Although non-combatants, we showed bravery under fire, resourcefulness, persistence, and a passion to aid suffering that outranks other emotions. Always woefully short of personnel – I here use the term to designate officers, enlisted men, and nurses – we somehow managed to care for all of our wounded, from the first-aid stations to the hospital ships which brought them home. Permit me to quote from the Surgeon General's annual report for the year 1919. On July 27th, 1918, he cabled Washington that: "The recent fighting has been so severe that the resources of the Medical Department have been practically exhausted in so far as personnel is concerned." He adds in the same report that: "The situation was saved only by the self-sacrificing spirit of officers, nurses and men. Officers, nurses and men worked themselves to the limit of physical endurance and that limit was beyond any which might be expected of human beings!" In this same report, General Ireland devotes a sub-chapter as tribute to the medical department men and women from civil life. I quote: "The majority of regular army medical officers called to the theater of operations, even if they had not had previous war experience, at least through their theoretical instruction and long association with trained officers, possessed some comprehensive idea of the difficulties and trials they were soon to face in Europe. However, with the reservist, the citizen-soldier – it was a radically different proposition. The sudden change to a primitive and unfamiliar environment opened up to him a field of endeavor strewn with obstacles of which he could not possibly have had the remotest conception. To these men of the National Guard, of the National Army and of the Reserve, to those front-line medical men, be they renowned surgeons or humble ambulance drivers, or litter-bearers, we desire to pay a well merited and special tribute to their loyalty, cheerfulness, initiative, adaptability and self-sacrificing devotion to the duties they shouldered in coming overseas. It's a far cry from a well appointed and comfortable physician's office in any bustling city back in America, with orderly routine the keynote of one's daily vocation, to the poorly lighted, unheated cellar of a battle-scarred and shell-menaced remnant of a building on a French battle-field, coping with emergencies that at times must have seemed insuperable. The skill and promptitude with which they succored the wounded reflected great credit on the profession they had the honor to represent. That they could so quickly adjust themselves to the work at hand, maintain an endless optimism, and render such splendid service with the meagre conveniences provided them was marvelous, and will always remain indelibly impressed on our memory as one of the brightest pages in the medical chronicles of the war." That the efforts and accomplishments of the Medical, Dental, and Nursing corps were appreciated by our superior officers as well as by the War Department in Washington is attested by the fact that, out of some thirty-five thousand officers of our service who donned uniform and followed the stars and stripes, 88 received the Distinguished Service Cross and 150 the Distinguished Service Medal. Medical officers received 5.1% of the total number of D.S.C.s conferred and 13% of the total D.S.M.s conferred in the Army. I wish time permitted reading more of these citations - you will find them all in a small book entitled *The Medical book of Merit*, published in 1925 by the Association of Military Surgeons in Washington.

Prevention and preparedness are, as perhaps never before in the history of our country, the order of the day. We see all around us, in almost every branch of human energy and human need, an earnest thought for the future welfare of all of us. I need hardly remind you of the magnificent progress the Illinois State Department of Public Welfare has made and is still making in relation to the broader aspect of medicine. The splendid group of buildings immediately to the south of this auditorium will, when fully completed and occupied, be a constant reminder that prevention and preparedness are the pillars of Hercules upon which the welfare of State and Country must rest. We stand committed to a campaign having for it's ultimate object the prevention - not only the cure - of physical conflagrations which might damage or destroy our well-being, and we are erecting adequate defenses against any and all potential bacterial, degenerative, or neoplastic enemies. Is it stretching our imagination too far to advocate and practice the same spirit of preparedness for our military and naval forces? Back in the days of the Roman Empire, the dictum: "Silvis pacem, para bellum" – if you desire peace, prepare for war, maintained a relative peace throughout the then known world - until disintegrating forces caused it's decline and fall. Human nature has not changed through the centuries; the struggle for existence still continues, with the individual as with the nation. Some of you, who may know little or nothing of the intricacies of military science and tactics, may perhaps ask yourselves: "why prepare for defense when we are at peace with the world; why not wait until we are at least threatened with war?" Would you advocate waiting until there has been a disastrous fire in town before buying fire-engines or training your hook-and-ladder companies in the art of fire-fighting; would you wait until a thief, murderer, or rapist has broken into your home before buying a lock and key for your doors and latches for your windows; would the farmer wait until lighting has struck and destroyed his barn before buying and setting up lightning conductors; did you married ladies – or would you future brides – wait until the wedding bells had ceased ringing before taking your first cooking lessons? If you did – or are planning to do so, God help your dispeptic homes! Every activity, individual or collective, is associated with preparedness – yet our pacifists and parlor-pinks would have us believe that preparedness is all wrong, that it brings on war. If war between nations is unthinkable – according to some of our myopic fellow-citizens – how about war between individuals? Why not be logical and begin on a small scale, just to see how this desirable Utopia might work out? Let us abolish the city police, remove the door-locks and window catches, and see how the Golden Rule is observed. If you cannot trust all the people in your community, if you believe in preparedness against burglary, murder, or rape; if you lock doors and windows by night and use door-chains by day as a means of defense against the greed, dishonesty, or degeneracy of some of your fellow-citizens, then, in God's name, come out and champion the same preparedness and means of defense for your own country and mine! Theodore Roosevelt had preparedness in mind when he quoted the old adage "speak softly but carry a big stick." The strong man is not necessarily a bully seeking fights in order to show off his strength; more often than not, he is gentle of manner and tolerant of his smaller brother's weaknesses. So with our nation conscious of our strength and permanent state of preparedness to resist any encroachment upon our sovereignty, we will continue to act as a deterrent to inordinate greed or covetuousness on the part of other nations. I have already spoken about our woeful unpreparedness in 1917; it almost makes one shiver to think what would have happened to our military forces had we then been attached by a first-class power – and not had Allies in front of us to hold the battle-lines until we were better trained, and to supply us with the war materials we lacked! We are not imperialistic; we already possess more than we can at present use; our yet undeveloped resources are sufficient to occupy the genius and energy of our people for generations yet to come. We merely want to keep what we have; our army and navy are the door-lock and window-catch which you and I know are necessary for our safety in any large community. Stand up therefore for preparedness; speak and act in favor of the regular army, navy, national guard, and organized reserves; they are your national police force! Join the guard or reserves, learn the rudiments of military science, do your patriotic bit. Are not many of the disturbing factors and events going on in Europe to-day strangely reminiscent of the bluster and sword rattling which preceeded the world war of 1914 – 1918? Should we again become the target of provocation after provocation, do you think we, as a virile nation, would long continue to be "too proud to fight!"

George de Tarnowsky
Colonel M.C. (Res)
Armistice Day 1938
University of Illinois School of Medicine.

Wartime Mission

1917 – 1918

Miscellaneous Writings

Traveling to Paris & Chateau d'Annel

October 1917

What of the Chateau d'Annel: how did it come to be turned into a hospital for French wounded; why was there a mixed staff of French and American surgeons? A wealthy expatriate American woman was living in the chateau when war broke out in August 1914.

Traveling to Paris

...the entire British nation was on heroic rations! The people themselves, civilians, military, and naval personnel looked anxious, sober-faced, pensive. That they were carrying on with that bulldog tenacity which has always been so characteristic of the British was self-evident; that the terrific loss of ships resulting from the ruthless submarine warfare which the Germans had been waging for many months had depleted England's food supply to a dangerous low level could not be denied. It might be said that the British stomach was "back to the wall" in the late months of 1917!

Colonel Scott, as senior Line Officer, had been handed a "travel warrant" for the entire group of American officers. He had previously hinted to me that he would like to stop off in Liverpool long enough to visit some friends of his. Just as we were leaving the docks for our afternoon tea stroll, he handed me the document with the parting words: "don't wait for me; I may not turn up!" He didn't; we never saw him again! The rest of us obediently boarded the afternoon train for London. Just as we emerged from our compartments in the metropolis, an air raid was announced: the dreaded Zeppelins were circling somewhere overhead! A London Bobby rapidly herded us all in the "Penny Tube" as London's oldest subway is called and stood guard over us for some two hours when, the "all clear" signal having been received, we were allowed to get into the night train for Southhampton. This, our first experience with wartime train service, gave a small inkling of the difficulties and uncertainties of travel. Three times did our train leave the London station; three times was it hurriedly signaled back before it actually was on it's way. Switchings, detours, hourly inspections of our credentials by members of the British Intelligence service made sleep all but impossible. It was a worn out, bedraggled lot of officers who finally marched to the "rest camp" which the Army had installed on a hillside overlooking the port. We were given a scanty breakfast - the classical herring, toast, marmalade, and tea - and then granted permission to wander about

Southhampton but to return sundown to Camp as we might get sailing orders for Le Havre "if the bloody Submarines keep out of the way", as the sargeant-major in charge of us put it! None of us had any Sam Browne belts, and we were advised to purchase them before sailing for France. The stores seemed to have unlimited quantities of leather goods on sale, and many of us added to our Field equipment that day. At ten o'clock that same night, just as we were ready to turn in for a well-earned night's rest, orders came directing us to pack up our equipment and march, under escort, to the Southhampton-Le Havre docks. A hazy moon was shimmering overhead, and a light fog hid all shipping from view. No lights in streets or houses: total darkness on board steamer. Like a flock of bewildered sheep, we followed our escort, trucks - or motor lorries as they were called by the British - carrying our equipment for us. The latter was piled up on the stern upper deck, covered with tarpaulin, and placed under guard. We were given the priviledge of either staying on deck with life preservers on or of occupying cabin berths. Discussing our chances if the boat was torpedoed in the English channel, we decided that we would drown anyway, on deck or below decks, so why not get some sleep - even if it turned out to be a mere preliminary to an eternal one? We asked for and were given a good cabin with two berths, undressed, and actually slept soundly until the steamer slipped up to dock at Le Havre.

"Abri des Permissionnaires"

France at last! Many of the younger officers who had remained on deck all night presented a forlorn appearance, dishevelled, blear-eyed from lack of sleep, and cold from the fog which had penetrated their uniforms. But were they happy to find themselves landing on terra-firma! Le Havre having been given over to the British Expeditionary Forces as a base, we were met by an English officer who had instructions to conduct us to Rest Camp No 2 on the outskirts of the city. Rest Camps, during the War, were designated places where soldiers on furlough or officers on leave of absence could get a meal, a bed, and shelter while awaiting transportation by rail or boat as the case might be. In the zone of the French armies, they were called "Abri des Permissionnaires" - it so happened that I was destined to use them frequently during my wanderings across the extended Western Front. Having shown our military papers and registered, we were free to wander about town for the rest of the day. Curious to see how France had stood the past thirty-four months of war (we landed on September 30th), I decided to "browse" about alone in order to talk to civilians and find out what their thoughts really were. Hopping on a trolley car gave me a striking illustration of the immense number of men engaged in defending their beloved Patrie! There were no motormen, no conductors on the trolleys, both positions being filled by women. The conductress of the trolley I rode on was a comely young woman in her early thirties wearing a plain black uniform devoid of all ornamentation. On inquiry, she informed me that three brothers had been called to the colors, two of whom had been killed during the tragic retreat which preceded the glorious first battle of the Marne. Her husband was also under arms but "Grace a Dieu et a la Sainte Vierge Marie" - thanks to God and the blessed Virgin - he had not been wounded, so far. A group of British Tommies got on board at one of the regular stops. One of them, speaking French fairly well and quite evidently under the influence of liquor, made a somewhat aggressive remark to the young woman. Without raising her voice or manifesting any justifiable anger, she merely lifted a medallion which she wore on her breast, containing a photograph of a French soldier, and said: "Mon mari, Monsieur: il est au front et je le remplace". My husband, Sir; he is at the front, and I am taking his place! What a magnificent answer to make! I almost clapped my hands in glee. The soldier had a decent side to him, after all, for he hurriedly murmured "beaucoup pardon, Madame", hopped off the car, and disappeared into the crowd. Lunch in a restaurant gave one another insight into France's difficulties. The meat was evidently horse's meat - even the famous French sauces could not disguise it's toughness and strong taste -, vegetables were scarce, and no bread could be had without a bread-card - which I did not possess. A French officer sitting at the table next to mine politely offered me the use of his own card; I evened the score by asking him to share my bottle of wine and to join me in the demi-tasse without which no French meal is considered a meal. Alas! Chicory has always been mixed with coffee in France, but the concoction brought to us was certainly virgin of the roasted bean; I judged it to be a mixture of roast barley and a small, very small amount of chicory. Sugar was non-existant; in it's stead were decanters of saccharin, the combination was beyond description but I bravely swallowed the mixture. My French friend, who had been observing me quizzically, smiled, shrugged his shoulders, and murmured "C'est la Guerre". It is War! How many, many times was I to hear the same expression, to see the same shrug of shoulders! It settled everything, from lice to delayed trains, from trench-foot to bombs; everything was "c'est la guerre"! Thus was brave France able to carry on; gay, irresponsible, pleasure loving France almost overnight acquired nerves of steel and a capacity for unflinching punishment which, as time goes by, will be extolled by historians. "C'est la Guerre"!

Of our night spent in Rest Camp No 2, the less written the better. The climate of Le Havre is naturally damp, rains had been a daily occurence for over a month, and the camp was actually soggy - cots, bedding, floors, and tents - nothing was dry. Before retiring, each officer had to sign a notice which stated that we were to be awakened at half past two on the following morning, baggage to be ready at 3.15, breakfast at 3.30, and departure from camp for the railway station at 4.A.M. Our destination was Paris where we were to report at temporary Headquarters, U.S. Army, then located at No 10 rue Ste Anne. The British provided a guide to the station and lorries for our luggage. From somewhere appeared an American corporal with his squad, carrying a case of hard tack, canned beef, and cheese for our journey. So far, we had not seen a single American Officer, but our British friends were looking after us quite satisfactorily.

The only excitement encountered during our twelve hour journey to Paris occured when our train was put onto a siding in order to let a British troop train go by. On another siding was a local train, also waiting for a clear track. A number of French women leaned out of the car windows and, unfamiliar with our uniform, one of them called out to her companions - in French, of course – “look, look! German prisoners!” I leaned out of my car window and shouted - also in French, of course - "No, no, we are American officers!" Laughter, clapping of hands, kisses politely blown across the interval between the two sidings, and “Vive les Américains!” charmingly ended the episode.

Battledore and Shuttlecock!

When I was requested to and accepted the task of writing War Manual No 7., my ignorance of military jealousies and intrigues led me to accept the mission as embodying the unanimous desire of the Medical Corps. When the editorial committee of the medical section for the Council on National Defense accepted my preliminary notes and wrote to the Surgeon General for specific orders, which would enable me to finish the book after visiting the front lines, the whole matter seemed settled. It was only when I read my general orders for overseas duty - in which my special mission was not even mentioned – that I became uneasy in my mind. In spite of assurances to the contrary, I sailed with misgivings; there seemed to be a catch somewhere but - orders are orders!

Arriving in Paris late in the afternoon of October third, 1917, I piloted my travelling companions to a quiet little hotel near the Madeleine called Hotel de L'Arcade in the street of the same name. After a grand night's sleep, I presented myself at No 10 rue Ste Anne, near the Opera. The building was a hotel turned into Headquarters, Lines of Communication or, as it would now be termed, Service of Supplies. The chief surgeon, L. of C. welcomed me freezingly; in fact, he so lacked in ordinary cordiality that I purposely refrain to mention his name. Regardless of where the trouble lay between the Surgeon General's office in

Washington and the Chief Surgeon A.E.F., all must have known that a Reserve Officer was merely trying to carry out orders to the best of his ability and couldn't possibly have a hand in any military camarilla. The peppery Colonel thought otherwise: "I know nothing about any book on military surgery you are supposed to be writing; I don't know why you were sent over for that purpose" was the burden of his remarks to me. A copy of the letter written to the Surgeon General by the editorial committee which I showed him merely elicited the growl: "This is no order; it's not even signed!" In desperation, I asked him if he would give me letters of introduction to proper French authorities so that I might get additional information from them. This seemed to be the last straw: "You cannot get any such letters from this office, and you are not authorized to represent yourself to the French authorities as having any mission connected with the Medical Corps. Leave your Paris address with one of the clerks and report back daily for further orders." So utterly unfair had been my treatment at the Colonel's hands, so uncalled for his trying to make me the "goat" of some, to me, unknown cross-currents in the higher staff that, instead of a feeling of defeat, my determination to finish the job was only intensified. A brother-in-law of mine was a member of the French Parliament, and through him, I soon received the following open-sesame to French military hospitals:

♦♦♦♦♦

Republique Francaise.

Ministere de la Guerre — Paris le 13 Oct.1917

Headquarters Lines Of Communication,
American Expeditionary Forces.

Special order. — France, October 17th, 1917
No 70

Extract

* * * * * * * * * * *

Par.3. Major George de Tarnowsky, Medical Reserve Corps, having been assigned to Lines of Communication by Par.21, S.O.125, A.E.F., October 13th, 1917, will proceed to Base Hospital No 6, reporting upon arrival to the Commanding Officer for duty.

Travel directed is necessary in military service.

* * * * * * * * * * *

By Command of Major General Blatchford

J.P. McAdams
Major, Inf. U.S.A.

Official:
Orrin A. Wolfe
Adjutant General

♦♦♦♦♦

Base Hospital No 6 was the Massachussetts General Hospital unit, located at Bordeaux, almost the farthest point from the battle lines. The order to report there could mean only one thing; the Medical Corps in France was getting rid of me and making my mission impossible. Bordeaux was actually not far from the Spanish frontier; what chance could I possibly have to study the giving of First Aid and transportation of wounded? This new move seemed indeed the final blow, but one last protest against the order remained open and I seized it at once. Requesting the Adjutant to furnish me with a Field Clerk from his office, I dictated a letter to the Surgeon General in Washington, quoting my orders for Bordeaux and explaining to him how, under the circumstances, it would be impossible for me to finish my text book. I asked for six copies of this letter, sent two "through channels" to Washington, and personally mailed extra copies to the Surgeon General and the Council on National Defense. The Adjutant looked daggers, but he well knew that letters, written through channels would have to reach Washington.

It was with a sad heart that I packed up my belongings, paid my hotel bill, and caught the night train for Bordeaux and - as it looked to me then - military oblivion!

Base hospital No 6 had taken over a large college building in Talence, a suburb of Bordeaux. Many of the surgeons were old friends or acquaintances of mine, and I was made to feel thoroughly at home. Nobody could understand why I had been ordered to the Massachussetts General Hospital unit as there was absolutely no surgical work expected for many months, and the personnel was busy rearranging the buildings in order to readapt them for hospital use. I wandered around Bordeaux, tried to do a little writing, and generally wasted my time, fretting at the bit yet unable to make a move.

Five days later, a telegram arrived from General Bradley, Chief Surgeon, American Expeditionary Forces, relieving me from my present assignment. The order was as follows:

♦♦♦♦♦

From HAEF - Oct 23/17.

Commanding Officer Base Hospital No 6 AEF Bordeaux.

Paragraph twenty-one, special orders one thirty four these headquarters directs Major de Tarnowsky to proceed to Annel Hospital for temporary duty. His order will be with Major Lambert, Chief Surgeon American Red Cross to whom he should report to make necessary arrangement to get into the zone of the army. Acknowledge.

Bradley.

♦♦♦♦♦

Channeling Back on Mission

My good colleagues, who knew of my plight and sympathised with a Reserve Officer's difficulties, were frankly delighted at the happy outcome and congratulated me on my good luck. It did not take me long to pack up once more and reach Paris. The American Red Cross had sponsored and supplied various hospitals behind the French lines, among them the Ambulance des Alliés in the Chateau d'Annel between the towns of Compiegne and Ribecourt and close to the Oise river. Both French and American Army surgeons were stationed there, and there had been some friction due, in large part, to a lack of knowledge of both languages. Besides taking over the duties of Chief Surgeon, I was expected to smooth matters over and obtain more cordial cooperation between hot headed young Frenchmen and equally hot-headed and superiority-complexed Americans! Much as I have always liked the French, I must admit that, professionally at least, they are absolute slaves to tradition, clinging blindly to whatever former leaders had once taught was the proper procedure in any given case. Seldom do French surgeons accept improvements which do not emanate from one of their own clinics.

After over three years of war, French surgeons had evolved a technique of wound treatment which they considered the last word in military surgery. The slightest deviation from this elaborate technique evoqued bitter criticism on their part. Many of the details of wound dressing were unnecessary but - they had been taught that way, ergo that was the one and only way! As all of the wounded at that time were French, one could not help but sympathise with their viewpoint, but our young American Surgeons did not propose to be told what to do by "Froggies" who couldn't even talk English! I finally had to tell my comrades that, if they could not conform to the French technique, they would be sent back to their own units to wait for months before they had any professional work to do. After this ultimatum, there was comparative peace but little cordiality between the two races. If a wound became infected - as so many war wounds necessarily did, and one of my surgeons had attended the case, one of the French surgeons would slyly write at the bottom of the surgical sheet: "Wound treated by systeme Américain; infected, naturellement!" This remark would create a furor for a day or two, and then peace would reign once more for a few days. Physically, my officers were giants compared to their French colleagues; there was little danger of any physical encounter, but the sly, anonymous stabs kept reappearing. The best argument with my own officers was to ask them to - in their imagination - turn the tables. How would they feel if the wounded were Americans treated by French surgeons? Would they not also insist on having our own boys given the best surgical treatment, according to American standards of surgery? As a matter of historic record, in July 1918, the French Medical Corps agreed to treat our own wounded because we had no facilities of our own. The result at Crépy-en-Valois raised an acrid controversy and from that moment to the Armistice, our wounded were exclusively treated by our own Medical Officers!

Chateau d'Annel

What of the Chateau d'Annel: how did it come to be turned into a hospital for French wounded; why was there a mixed staff of French and American surgeons? A wealthy expatriate American woman was living in the chateau when war broke out in August 1914. Through friends in New-York and the American Red Cross, she opened up a small private hospital to which were sent mostly convalescent soldiers in need of prolonged hospitalization. When we entered the war, the Red Cross thought fit to enlarge the scope of this hospital and build around the chateau a number of wooden wards called "barraques Adrian" by the French. The new hospital was given the name of Ambulance des Alliés, Ambulance being a term used in France to denote any mobile or semi-mobile hospital near the front lines. Being henceforth sponsored by the American Red Cross, it came under the direct jurisdiction of Major Lambert, and through him was alloted the American staff of surgeons whose status was that of U.S. Army Surgeons on detached duty with the Red Cross. Like most of the chateaus of the Louis XIV period, the one at d'Annel consisted of a main building with two wings and a courtyard. Two wards, with the operating room, X ray, and laboratory had been installed on the main floor, leaving the parlors and dining room for use of the family and staff. The second floor was reserved for sleeping quarters for family and senior officers. The junior officers, nurses, and soldiers had quarters in barracks. The commanding officer of the unit was Major Cassan, a French officer of the regular army and a veteran of many colonial campaigns. He was a dapper little chap who couldn't possibly have weighed more than one hundred and ten pounds or stood higher than five foot four without rising on tip-toes. With his bristling mustache, jaunty quick step, and habit of turning his head from side to side as he walked, he forcibly reminded one of a chipmunk. His conversation was as rapid as his walk, and his habit of prefacing every other sentence with the French equivalent of "you know" made us Americans dub him "Monsieur n'est-ce-pas!" Major Cassan was as putty in the hands of Madame la Chatelaine and obeyed her slightest whim.

The hospital orderlies were all French reservists, elderly men in their fifties inducted into service because poor France had by that time been almost "bled white", as the Germans proclaimed they would do in order to end the war. These old "Pépères" - Fathers - as they were called in the army, had had no previous hospital training and were merely useful as stretcher-bearers and scrub-men. Among them were a number of priests who had volunteered for service. On the chateau grounds was a small chapel, and it was no unusual sight to find one of these devoted patriots holding early mass. All wore military uniform, and it was only by recognizing them at their matins that their true identity was revealed. One of them, Father Bonne, had an outstanding personality and a brain teeming with general knowledge. Presumably he had been educated in a Jesuit college, though he never told me much about himself. When a test of courage came in March 1918, Father Bonne was the only orderly who asked and was granted permission to remain at the hospital when it was ordered evacuated. But of this episode, more anon!

The nursing force consisted of a few graduate British nurses and many volunteer American girls who had taken this opportunity to see War at first hand and do their unselfish, heroic bit. Three nurses from the Roosevelt Hospital of New-York-City formed the operating room team, a most efficient trio. All shades of social background and intellectual attainment - from Bryn-Mawr graduates like E.L. and R.W. (the inseparables, we called them) to daughters of humble British homes where the habit of dropping "Haiche's" had not yet been lost - were represented in the nurse's quarters, all of them gathered around a central, dominant idea - rehabilitation of war victims. Presiding over the hospital - one of the many amateur hospitals which had sprung up in 1914 as a gesture of good-will towards France and had been classed by military authorities as "Hopitaux Bénévoles" - was the "Directrice", the aforementioned New-York expatriate. Her exact status, aside from the fact that the chateau was her home, was a very indeterminate one. To use a familiar military term, she was neither accountable nor responsible for anything which went on in the hospital yet, by virtue of courtesy more than anything else, she dictated the hospital's policies. She did have the hiring and firing of the nursing staff, because these salaries were paid either by herself or by the American Red Cross. The Administrative functions were entirely in the hands of the French Medical Corps. What the Red Cross or the Directrice supplied in the way of clothing or supplementary food was in the form of donations. This situation must be understood because it explains, in part at least, how a military hospital could be evacuated without higher authority by a panicky American woman who feared possible capture!

The chatelaine had her own suite of rooms in one of the chateau wings; her daughter a similar suite in the opposite wing. The civilian servants were such as one would expect to find in a well chateau. Two other members of the household stand out as intellectual personalities: the housekeeper a gentle white-haired refined lady who manifestly occupied her present humble position through recent financial reverses, and an Alsatian governess or language teacher who had evidently been in the family service for many years. She was patriotism personified: her love of France and hatred of Germany equally passionate; she would weep buckets of tears whenever her beloved "Poilus" the hairy ones, as French soldiers were affectionately called, all because, during the hectic weeks preceding the first battle of the Marne they had been too busy fighting and retreating to shave, were beaten back anywhere along the battle line. Correspondingly, she would "march on to Berlin" if the Germans received a thrashing and talk jubilantly about the day when Alsace-Lorraine became, once more French territory.

Many French Colonels and Generals had their entrée at the Chateau d'Annel, and elaborate banquets were of common occurence. We surgeons always shared the family meals, paying our Mess bills to the Majordomo who charged us a fixed price per day according to rank - the higher the rank, the higher the bill. The Directrice was, at that time, a Divorcée, and it became quite apparent that she did not lack in potential suitors. One General in particular, in command of the Army Corps fronting our sector, was most attentive and solicitous - Madame later married him! [Sometime later with the Madame's generosity, the General] promissed to give me the opportunity I so eagerly sought of visiting the front lines and studying the French method of treating war wounds.

It will always be a great source of satisfaction to me to reflect that the continuation of my work was made possible through fortuitous circumstances - or Lady Luck - and that only later did the Surgeon General and Chief Surgeon, American Expeditionary Forces come to an agreement regarding the value of the information I was gathering for the benefit of our Army Surgeons. Believe it or not, as the cartoonist Ripley says, personal glory never entered in any of my calculations. Entering the Military Service in a spirit of pure patriotism, deeply anxious to show my devotion to the flag of my country of adoption, every mission entrusted to me was carried out to the very best of my ability. I gave all of me then to the service of the United States and felt deeply and humbly grateful when official recognition of my efforts crowned what, to me, will ever be the great romance of my life. Whatever bitterness my early experiences may have brought, time has completely mellowed [me], and I feel no grudge against the few superior officers of the Medical Corps who seemed to be determined not to let me succeed. That they were working off some personal disappointment, some actual grudge, through me, I am convinced. Many details of this military cabale have trickled through, but why revive them now? The loss of valuable time was the worst feature of this underground quarrel of which I was made the temporary victim. My visits to the front lines had been limited to four weeks. Arriving in Paris on October third, it was only ten days later that the French authorities sent me their authorization to visit their military hospitals, and November 12th, 1917, as guest of General T... I actually penetrated French trenches and began to visualize what we had to learn if we were to become competent military surgeons.

As it is doubtful that armies will ever again come to such a complete stalemate as that of the World War in 1917, so is it highly improbable that such elaborate systems of trenches and dugouts will ever be seen again. From an historical view-point, it seems worthwhile to copy, verbatim, the extraordinary methods employed in order to lessen casualties. As the topography of each sector was different from many of the others, certain types were described in my book. The French Zone of the Advance represented every kind of natural obstacle, whereas the British lines in Flanders were solely concerned with plains, mud, and water. In 1917, the censor could not permit any mention of towns, valleys, or hills; these have been added to my original notes. As fast as each inspection was completed, original chapters, begun in Washington, were revised and brought to date before mailing them to the Surgeon General's office. By immediately issuing thousands of mimeographed copies of my chapters and sending them to the various camps where Medical officers were receiving their training before being sent to France, the serious loss of time was, to some extent, overcome. The book itself was not ready for sale or distribution before June 1918, nine months after my arrival in France!

Miscellaneous Letters

Orders Regarding Book on War Surgery

November 1917

The purport of the letter was that Major de Tarnowsky had been designated by the Editorial Committee, Council of National Defense, to prepare a book on war surgery.

The Orders

That my letter to General Gorgas, quoting my orders to proceed to Bordeaux and report to the C.O. of Base Hospital No 6, soon bore fruit was most gratifying to me. On November 13th, the following order was handed to me when, "through channels" I mailed a few of the revised chapters of the book and left a copy of them with Colonel W......, Surgeon, Service of Supplies.

♦♦♦♦♦

Headquarters Lines of Communication.
American Expeditionary Forces.
France.

November 13, 1917.

Major George de Tarnowsky, M.R.C. has permission to proceed from Paris to Chaumont and return.

(Signed) Orrin R. Wolfe
Orrin R. Wolfe
Adjutant General.

♦♦♦♦♦

Notice however that I was merely granted "permission" and that the travel was not considered "necessary in the military service" – which meant that I had to pay my expenses both ways! To be able to talk to the Chief Surgeon was a step in advance. General Pershing and staff had been given a large public building in Chaumont for G.H.Q.: General Bradley and staff occupied one of the wings, and I was soon closeted with a group of senior officers who, while not exuding cordiality, nevertheless agreed to give me the opportunities I sought. Returning to Paris that same night, too late to reach the Ambulance des Alliés, I wandered from hotel to hotel in search of a room. Finally, in despair, I spied a squalid hotel across the front of which had been nailed a sign stating that it was managed by the Y.M.C.A. Entering, the night clerk grumblingly informed me that there was a vacant bed on the third floor which was still empty. Accepting the offer, I picked up my kit bag and started upstairs. "Wait a minute, wait a minute" said the clerk; "you can't go upstairs without paying in advance!" Too tired to enter into a religious discussion at that early hour of the morning, I paid, found the room and bed, and crawled in, utterly exhausted. Alas! many other living creatures had crawled in ahead of me and, resenting an intruder, proceeded to attack various and sundry parts of my anatomy! I stood the torture for a half hour, turned on the light, exterminated all of the vermin I could see or feel, dressed, and walked out on the streets of Paris. Reaching the Gare du Nord, the night watchman allowed me to sit on a bench and thus, with my kit bag clasped around my arms, passed the remainder of a most wretched night. This, and subsequent war experiences with "Y" men and "Y" activities, considerably lessened whatever esteem I may have once had for this organization. In all probability, lack of proper selection of the men sent over to France and Italy had much to do with the dubious reputation they left behind them – dubious regarding mercenary methods and un-natural morals.

On November 14th, 1917, the following order came through the mails:

♦♦♦♦♦

Headquarters
American Expeditionary Forces
Office of the Chief Surgeon

France, November 14, 1917.

From: Chief Surgeon

To: Major George de Tarnowsky, Ambulance des Allies, Chateau d'Annel, Secteur Postal 205.

Subject: Authorization to visit French Sanitary formations.

1. In order to enable you to carry out the instructions received by you in letter form, prior to your departure from the United States, you are hereby authorized to visit any French Sanitary formations which may be within the Zone of your activities, both in the Service of the Advance and Service of the Interior.

2. It is, however, directed that you will defer visits to the French Sanitary formations on the Zone of the Armies until such time as regular permission has been accorded you which has this day been requested through the French Mission at these headquarters.

3. As soon as this authorization has been received it will be furnished you as the credentials to be presented when you visit any of the formations above mentioned.

(signed) A.E. Bradley
A.E. Bradley
Brig. Gen., N.A.

♦♦♦♦♦

Events now began to move rapidly. Working at top speed, as fast as I returned from one of my tours of inspection chapters would be revised and sent on to Washington. A copy was always mailed to the Chief Surgeon. On December 2, 1917, the Surgeon General in Washington wrote the following letter to General Bradley:

♦♦♦♦♦

(Copy)

December 2, 1917.

From: The Surgeon General
To: The Chief Surgeon,
American Expeditionary Forces,
France.
Subject: Major George de Tarnowsky, Medical Reserve Corps.

1. Major George de Tarnowsky, Medical Reserve Corps, was sent to Europe and ordered to report to the commanding general, American Expeditionary Forces, for assignment to duty. At that time, at the direction of the Surgeon General, letter was written explaining the status of Major de Tarnowsky. The purport of the letter was that Major de Tarnowsky had been designated by the Editorial Committee, Council of National Defense, to prepare a book on war surgery. The various chapters of this book were outlined in the United States from data at hand. It was felt that a considerable amount of revision might be necessary and your office was requested to aid Major de Tarnowsky in every possible way to

come in contact with the proper persons to enable him to revise such chapters of his book as might be necessary.

2. This office has received several letters from persons connected with the Council of National Defense and others, that Major de Tarnowsky has been unable to complete his work for the reason that he has been prevented from making the best use of material at hand and visiting those points at which material for his book was available.

3. It is believed that the book in preparation would be most valuable to the department, but it's value will be lessened by delay in completion. The Surgeon General, therefore, requests that everything possible be done to expedite the completion of this book by such assistance to Major de Tarnowsky as may be possible. Upon completion of this work Major de Tarnowsky will be assigned to such duty as the interests of the service may demand.

(signed) Robt E. Noble
Lieut. Col. Medical Corps.

♦♦♦♦♦

In rapid succession arrived formal military authorizations to visit the front lines in the St Quentin, Noyon, Soissons, Toul, and Verdun sectors. From three to six days were spent in each sector, sleeping in dug-outs, tents, cellars of bombarded houses, or a General's H.Q. As the French authorizations were all similarly worded, only one of them will suffice to give the reader an idea of the preciseness which had to be followed during the War.

♦♦♦♦♦

Le 21 Novembre 1917.

Mission Militaire Française
près l'Armée Américaine.

Secteur Postal 10.
Direction des Services.
Bureau
Santé
No 12.106

Le Colonel Reboul, directeur des services de la Mission Militaire Française près l'Armée Américaine.

a Monsieur le Général Commandant en chef A.E.F

Objet: Autorization pour le Major Georges de Tarnowsky de visiter les formations sanitaires Françaises.

Sante

Réponse a lettre No 210.715 du 14 Nov. 1917.
1496

J'ai l'honneur de vous faire connaitre que par lettre No 8058/DA du 19 Novembre 1917, le General Commandant en chef: "autorice le Major G. de Tarnowsky du corps medical Américain, actuellement en service a l'Ambulance des Alliés, a visiter les formations sanitaires de la 3ème Armée.

En arrivant dans cette armée, il devra se présenter a Noyon, au Chef Supérieur du Service de Santé qui réglera l'itineraire et fera accompagner Monsieur de Tarnowsky par un médecin militaire.

Le voyage n'excédera pas 3 jours" Conformément a votre désir, ce voyage commencera le 26 Novembre 1917.

(signed) Reboul.

♦♦♦♦♦

As soon as the chapters on the French Zone of the Advance had been written up and mailed, I asked for and promptly received the following orders:

♦♦♦♦♦

Headquarters
American Expeditionary Forces
office of the chief surgeon.

France, December 25, 1917.

From: Chief Surgeon, A.E.F.
To: Major George de Tarnowsky M.O.R.C.
Ambulance des Alliés.
Subject: Order Authorizing Procedure to 4th British Army, Cassel.

1. Inclosed herewith is an order authorizing you to proceed to the 4th Army and report to the D.D.M.S., 4th Army (located at Cassel).

2. The English Mission at these headquarters have made a special request that they be notified as to the date you expect to report at Cassel. Will you please telegraph us this information just as soon as you have determined the date, so that the Mission may be notified.

By direction of the Chief Surgeon:

(signed) M.W. Ireland
Merritte W. Ireland
Colonel, Medical Corps.

♦♦♦♦♦

(Cassel is a small town in Flanders. The D.D.M.S. referred to in the letter meant: Deputy Director Medical Services. Colonel Ireland, soon after dictating the letter to me, became Chief Surgeon, A.E.F. and later, Surgeon General.)

The enclosed order read as follows:

♦♦♦♦♦

Headquarters American Expeditionary Forces.
France.

December 24, 1917.

Special Orders
No 197

Extract

* * * * * * * * * * *

44.

Major George de Tarnowsky, Medical Reserve Corps will proceed to Headquarters, Fourth Army, British Expeditionary Forces, to carry out the instructions of the Commander-in-Chief, and upon completion of this duty will return to his proper station.

The travel directed is necessary in the military service.

By command of General Pershing
James G. Harbord
Brigadier General
Chief of Staff

Official:
Robert C. Davis
Adjutant General.

Finally Recognized

After almost three months spent in France, my mission was finally recognized as having some value for the Medical Corps. Henceforth, all my trips were to be "necessary in the military service", which meant that all expenses incurred were reimbursed me, in addition to my pay. As no stenographers were ever available, my faithful typewriter did Yeoman service. It was lucky that my room in the Chateau happened to be in an isolated part of the second floor, because it was nothing unusual for me to keep transcribing my notes until the wee hours of the morning. Luckily, the sector of the battlefield between Noyon and Soissons had become a faily quiet one, since the October battle of the Malmaison where American and French surgeons had jointly treated hundreds of wounded soldiers in mobile hospitals erected at Vauxrot for that particular offensive with "limited objective". There was, at that time, no thought of either side "breaking through", but both Germans and Allies sought to improve their defensive or offensive positions by nipping off salient or conquering a hill or two, which would give them better visibility. There would be a heavy admission of wounded for a day or two, followed by a week or ten days of relative calm. When the German communiqué said "nichts neues am Westen" or the French "Officiel" reported: "Rien a signaler" – nothing to note – it meant that only a few hundred dead or wounded represented the sum total of the day's activities, and no trenches had changed hands. Individual tragedies were not worth mentioning; only tactical advances or "strategic" retreats counted. One got so one could read between the lines and know pretty well what was going on, in spite of the diplomatic language used in all official releases. "According to plan, we retired to previously prepared lines of defense" meant that the Germans had lost a sector of battle-line; "We have consolidated our defense along new lines" meant that the French had had the worst of that particular engagement!

The Letters M.C.

On January 29th, 1918, I felt fully rewarded for my efforts on receiving the following letter:

♦♦♦♦♦

Headquarters
American Expeditionary Forces
Office of the Chief Surgeon.

January 29, 1918.

Major George de Tarnowsky, M.C.
Ambulance des Alliés.
Secteur Postal 205.
France.

My dear Major: -

Your report on Sanitary Service of the French zone of the Advance has recently been received in the office of the Chief Surgeon, A.E.F.. We have all read it with the greatest interest and General Bradley has asked me to acknowledge its receipt. This report with others on similar lines is being studied with a view to modify our equipment and organization in so far as is necessary to permit us to adopt in a general way the scheme which has been evolved by the French and which seems to work in a most satisfactory manner.

Let me express to you our appreciation, and with best wishes, I remain,

Sincerely yours,
(signed) S.H. Wadhams
Lieut. Colonel, M.C.

♦♦♦♦♦

The real joy-bringing part of the letter lay in the two capital letters M.C. after my name. It meant the official abolition of the word Reserve, which created so much ill feeling when we entered the war and which continued until we all became part of one army, the U.S. Army. Just why we should have been branded as Reserve and made to wear insignia proclaiming the fact is hard to explain: it certainly was poor psychology. Speaking for the Medical Corps, we were, by and large, just as well professionally trained as our colleagues of the regular army besides having clinical experience in civil surgery which the latter could not, in peace time, acquire because of lack of opportunity. Granted that we were, at first, somewhat deficient in knowledge of what is usually referred to as "Paper Work", i.e. orders, requisitions, sick and wounded reports, etc. which, after all was mostly clerical work easily learnt, and that we occasionally tripped over some "custom of the service", the real fact to retain was that we were sacrificing our civilian careers and giving the best that was in us for the good of the army. Civilian life ended with us in May 1917 when we volunteered for service and began again, quite at the bottom, some time in 1919 when we packed our war-stained uniforms in moth balls and looked around us for something to do! It has always seemed to me that the original friction in May 1917 occurred between the Regular Army and the National Guard, and that we, of the Reserve Corps became the innocent victims of the antagonism, which unquestionably existed between the two major components of the organized forces of the country. Was it due to a superiority complex on the part of the regulars or to an exaggerated sense of "State Sovereignty" and "chip on the shoulder" feeling on the part of the Guard?

In all of my tours of inspection and observation, I was received with open arms and given every opportunity to gather information. While the nights were far from pleasant and the underground sleeping quarters anything but dry and sanitary, the food was always good and conversation, aided and abetted by "Pinard" – as wine was universally known in the French army – flowed gaily and freely. No matter how hard they were pressed at times, the French officers and men I came in contact with rarely lost their native good humor, never seemed to despair of ultimate victory. The wanton destruction of trees, shrubs, and buildings carried out by the Germans at the time of the famous Hindenburg retreat when the Noyon salient was evacuated, steeled the French more than any other event of the war. Again and again was German psychology in error. One can well imagine the feelings which must have surged in the minds of the peasant soldiers as they contemplated the loss of thousands of fruit trees, sawed off a few feet from the ground, lying where they had fallen, mute examples of the "Schrecklichkeit" by means of which the Boche fondly and naively hoped to spare the enemy into submission!

Journal Studying French Bataillon

The Gentleman in a French Soldier

Tuesday, December 28, 1917

...such contrast with the British and French system in which the officer thinks of his men first and of himself last. I am constantly amazed at the beautiful democratic spirit, which prevails in the French army. Officers and men fraternize openly, and yet the discipline is excellent.

I have had most wonderful and thrilling experiences, studying the bataillon and regimental aid-stations in the region of France's most famous post. I was gone six days on my last trip and took one of my Lieutenants with me to take notes. Most of the time, we had a General's limousine at our disposal. I have received the most wonderful welcome everywhere and can never be grateful enough to the French officers, both Medical and Commanding, for their courtesy.

I visited a great many hospitals, including the wonderful one hewn out of the rocks under the famous Citadel. Some of the trips were extremely dangerous, and we had to take shelter under rocks or crouch down in the bottom trenches, while bullets and shells whizzed over our heads. When I tell you that I crawled up to a listening post and saw a German soldier fifteen yards off, in another peep-hole, blinking at me, you can judge of the nearness to the enemy lines. In many places, owing to swamps etc, no trenches can be built, and we had to dash across the open, one by one, in order not to draw fire. As a rule, they don't waste ammunition on a single soldier - but they have a nasty way of changing their minds without notifying you in advance. At another place, we had to turn back as they put a barrage fire in an open space ahead of us and we could not get through.

The only really hard time I had was in passing from one Army Zone to another. No automobiles were allowed to pass, so we had to take a train. We were to have been met at the post of -- by another auto, but the train was five hours late, and it was eleven o'clock at night before we reached the place. The auto had departed, leaving no instructions, so there we were, in a pitch dark town, under bombardment. We carried our suit cases with us and started out to find a hotel. All were packed to overflowing, so, after an hour's futile search, we tramped back to the station to spend the night in the cold dark waiting room. It was then after midnight and we hadn't eaten a thing since noon. The officer on guard then suggested that we try a rest home for soldiers on furlough, called "Abri des permission-aires" and showed us the way there. We found two bags of straw and two blankets apiece - the last ones. Through the same kind officer, the

Canteen or "Cooperative" was opened for us and I bought a box of cheese, a piece of sausage, a hunk of bread, and a bottle of wine which we ate at one o'clock in the morning. Going to bed was simplicity itself as we took off our shoes and put on an extra pair of heavy woolen socks.

Just as we had tucked ourselves in, and our teeth had begun to chatter from contact with the cold and very damp blankets and straw, two belated French officers came in and looked at us in despair. That they did not curse us simply goes to prove that politeness is an essential part of every Frenchman's makeup. They spent the night trying to keep up a small fire in the one tiny stove which the wooden barrack contained. C'est la guerre. In the morning, as all the water was frozen, we washed our faces with the corner of our handkerchief and reported at headquarters after drinking a big bowl of hot Cafe au lait.

There we were given another automobile and sallied forth to complete the trip. Over the furthest north of the cities of that region, we saw fourteen aeroplanes fighting over the town - a most thrilling sight. Two nights, we slept in officer's barracks in a big hospital center with wonderful underground passages where the wounded were carried during a bombardment; another night, I had a cot behind a little grocery store kept by two dear old peasant women. In my wanderings, I overheard two Niçois peasants (soldiers) talking. They turned six summersaults when I spoke to them in their own patois. All the regiment rushed up to see the wonderful American officer who not only spoke French but could talk patois. Had frozen feet for two weeks but am now well.

The Gentleman in a French Soldier

I have had six wonderful trips so far, visiting every part of the devastated district, talking to the people, walking through cities and villages where only the streets remain, climbing down into marvelous grottoes, caverns, and dug-outs made by the French and Germans, and studying the methods of picking up the wounded, carrying them to these underground rooms for first aid, and then transporting them by Auto, by rail, by boat, etc, to the various front hospitals - I have picked up a great collection of pieces of bombs, shells, grenades, and torpedoes.

In the course of my wanderings, I walked over what used to be [my nephew] Ivan's birthplace and have been in all the towns you saw in 1902. In all, I am sure, I have visited over a hundred towns and villages, always in a military auto, with a guide to get me in and out of the maze of trenches. The systematic devastation of the country is one of the things one cannot forgive the Boches for. Acres and acres of fruit trees chopped down, the trees lying where they fell, proving that the Boches did not need the wood, but only sought the devastation of a race. All girls of over 14 years taken away. One poor woman, with seven children, had her two daughters, 14 and 16 years old, taken away by the Boches. Her husband is a prisoner in Germany, and she was severely wounded in the abdomen by a shell. Her farm-house is half demolished, yet there she was, making a brave fight with her five young ones around her. We sent her bread, potatoes, and clothes and have given her two goats whose milk keeps the three youngest children alive. One dear old peasant woman kissed my hand and said "We thank the Americans for having come to aid us".

Her two daughters, both married, and three grand daughters have disappeared in Germany. In one village, not far from where you and I visited, two girls, 16 and 18 years old, were forced to submit to a medical examination in the presence of their father in order to see "if they would do" for the German soldiers.

And still, there are people in the U. S. who profess to wonder why we are at war with these Huns. I walked all over the secluded Chalet where the Kaiser's son Hubert had the best looking girls of the neighborhood brought to him at night-time. Over the front door, in large letters, are the words "Hubertus-Haus" and "Vivat Venus". All these things are not nice to write about, but we must know the truth. I did not believe these things before I had actually seen them and talked to the poor terror-stricken people who have had to live in Hell for the past three years.

The only time I feel like smiling, on one of these trips, is when I come across German prisoners rebuilding all the villages and roads they have destroyed. France is wisely making them, in a small way, undo the wrong they have done. Cemeteries have been systematically torn up, all the metal from the coffins removed, and the skeletons thrown around on the ground. Everywhere you see the arrogance and selfishness of the Prussian officers. They lived luxuriously in bomb proof houses or dug-outs, with thick iron and cement walls and roofs, while they locked up their poor soldiers in mitrailleuse cages so that they could not escape from shell fire. Always the best of everything for the officers and the worst of everything for their men; such a contrast with the British and French system in which the officer thinks of his men first and of himself last. I am constantly amazed at the beautiful democratic spirit which prevails in the French army. Officers and men fraternize openly, and yet the discipline is excellent.

The French soldier goes about his duties, because he knows that it is for the good of the service and of the country - not because he fears to be punished if he doesn't perform his duty.

I had a most interesting lunch a few days ago with General X—of X--th Army, in command of this sector. He was quartered, with his staff, in a small villa, fairly intact, living very simply in constant touch with his troops, by whom he was adored. The meal was excellent, and the conversation most interesting and highly amusing, as all French officers are learning English. They study two hours a day and quiz one another, but as they have no professor to give them the proper pronunciation, the result is often quite comical.

After lunch, the general took me in his own Auto for a hundred mile tour of his section, in and out all over the country. They all, wherever I go, marvel that an American should be able to speak French so perfectly and, as I am very tired of explaining the same old story over and over again, I just let them marvel.

The hospital organization is running very smoothly now, and everyone seems delighted to have me in charge. It certainly is a rare opportunity for the young Surgeons here, as otherwise they would all be twirling their thumbs in their Base Hospitals, waiting for our wounded men to begin coming in. Tell Mother I shall be glad to have the socks as the weather is very cold (as you can see by my hand writing), and two pairs of socks and two sweaters are necessary for comfort.

The British Army in Flanders Fields

Monday, December 24, 1917

Having thoroughly studied the French methods of treatment of war wounds, my next concern was to compare them with those of the British army, particularly because flat, muddy Flanders presented vastly different problems...The plan was to inspect the aid-posts, give me ample time to make notes on the treatment and evacuation of wounded, and return to H.Q. before the enemy laid down his daily barrage of shrapnel.

The British army in Flanders fields.

Headquarters American Expeditionary Force
France, December 24th 1917.

Special orders:

No 197

Extract.

44.

Major G...T...., Medical-Reserve-Corps, will proceed to Headquarters Fourth Army British Expeditionary Force, to carry out the instructions of Commander-in-chief, and upon completion of this duty will return to his proper station.

The travel directed is necessary in the military service.

By command of General Pershing
James G. Harbord
Brigadier General
Chief of Staff.

Official:
Robert C. Davis
Adjutant General.

Having thoroughly studied the French methods of treatment of war wounds, my next concern was to compare them with those of the British army, particularly because flat, muddy Flanders presented vastly different problems. In hilly terrain, trenches and dug-outs were simple tasks for the Engineer corps to solve, and first-aid could be rapidly given to wounded soldiers. The above order had been sent to me after permission had been granted to visit the British war zone, and I had notified them of the probable date of my arrival at H.Q. in Cassel, northern France.

Under the chapter of "General considerations", the official report of my tour of investigation contained the following statements: "It may be stated as a truism that any comparison between the French and English methods of treatment or evacuation of the wounded in their respective zones of advance is impossible, because of radical differences in the topography of the land and in the general conception of trench warfare.

The English have maintained a comparatively short front, difficult to hold and correspondingly more difficult to advance in. They have pinned their faith in density of front, with continuous activity along the entire front, alternating with intense attacks in selected sectors. The enemy, constantly and diligently harassed, retaliates in kind; the pressure is constantly maintained, night and day. This unending "strafing" has naturally multiplied the problems of evacuation, increased the distance between the front lines and regimental aid-posts, and lengthened the elapsed time between the reception of a wound and it's thorough surgical treatment. The evacuation hospitals or casualty clearing stations (C.C.S's) are under more or less constant bombardment, either by large caliber naval guns or aerial bombs; they have frequently had to be moved about from place to place and it has been impossible to keep the wounded in the zone of the lines of communication until their wounds are healed. The inevitable result has been that the British base hospitals have had to perform a good percentage of primary operations and a still larger percentage of secondary or completing operations.

Primary closure of wounds, after thorough excision of all devitalized tissues, which is now almost an axiom in the French army, is still in the experimental stage with the British (January 1918). This apparently radical difference in the zone of advance should not be laid at the door of conservatism; rather is it due to the time element, the extremely precarious position of the C.C.S's, and the consequent necessity of rapid evacuation of the wounded men to the base. Every effort is made to get the wounded to the C.C.S's while the wound infection is still localized; stretcher bearers and ambulance drivers are straining every effort towards expediting evacuations, working under surroundings of almost indescribable difficulties."

Reaching Cassel

In peace times, the trip from Paris to Cassel, H.Q. of the British fourth army, would have taken but a few hours by motor or train, but in January 1918, with the Germans in possession of St Quentin, Cambrai, and Lille, the only means of reaching the northermost French department (Département du Nord), was in a most roundabout way via Amiens, Boulogne, Calais, and Dunkerque. Four different trains carried one slowly, haltingly towards the battle-lines and, leaving Paris at eight in the morning, it was eight at night before the quaint, historic town of Cassel was reached. From Dunkerque on, the landscape had consisted of rain, fog, and mud. Intermittent flashes of light and the booming of artillery in the distance, with answering booms but no flashes of light from the British lines, ushered me in the war zone. So cleverly hidden were the British batteries that I never saw a single cannon during my entire tour of investigation; when visiting advance military formations in war-time, one naturally refrains from asking embarassing or suspicion-arousing questions!

My orders duly examined and stamped by the regulating officer, I found myself on the almost deserted station platform. Pitch darkness all around me; no lights, no matches or flash-lights allowed to be used. British sentinels suddenly looming out of the darkness, challenging one with their brisk "who goes there!". Military orders examined under cover of a doorway, vague advice to "walk straight up the road, sir, then turn to your left"; slipping, staggering in the slimy mud until a fairly large building loomed in front of me, and another sentinel snapped his challenge.

"Adjutant's office, yes, sir; this way, sir!"

Said the Adjutant: "How d'ye do, sir? Been expecting you sir, have a car for you in a short time; sit down, have a fag, and a little drink? Sorry we didn't know you were on the evening train sir; would have had a batman (orderly) for your traps." My "traps" consisted of musette and duffle-bag, both quite heavy and hard to carry; they were deposited on the office floor with a sigh of relief. A drink, a smoke, a shake-hands, and off I went in the D.D.M.S's (Deputy Director Medical Service) car to Caistres, a small nearby village. Formal greetings by various R.A.M.C. officers, another drink, another smoke, then to bed in the general's room, the latter being in "Blighty", i.e. England on leave. Instead of sleeping immediately, as nature imperatively demanded, I sat up in bed for the next three hours with a sputtering lamp, copying rules of instruction on "delousing" and "treatment of trench foot"! Looking back over the intervening years, one marvels at the sum of work a human being could accomplish under the nervous tension we were all living under! Small wonder is it that so many of the commanding generals only survived the Armistice a few years; after being taut through terrible months of anxiety, they snapped!

The Billy Goat

At breakfast the following morning, I was initiated into a most unique and efficient method of destroying confidential military messages. As the chief-of-staff took his place at the head of the table in the dining room of the villa we were occupying, a handful of despatches was handed to him by his orderly, who at the same time, solemly ushered in - a billy-goat! The latter stood - I almost want to say, at attention - at the colonel's right. Each despatch was opened, read, pondered over thoughtfully, crumpled up into a round ball, and handed over to the goat who promptly swallowed it and nonchalantly waited for more!

The unique meal over, Colonel G..., consulting surgeon of that army corps, took me in hand for a tour of inspection. Kemmel hill, St Jean, the Zonnebecke-paschendale ridge, and Messignes were explained to me, and my admiration went out to the dogged perseverance which alone had enabled our British comrades to hang on to the Ypres salient in the face of almost unsurmountable climatic and topographic difficulties. From three sides of the salient belched constant, harassing shell-fire; gas attacks were almost daily occurrences, and aerial bombs crashed on hospitals, munition dumps, railway tracks, and duck-boards indiscriminately. Mud, mud, mud everywhere, slimy, slippery, treacherous as quicksand. Dug-outs under roads with muddy, smelly water on the floors and dripping down through the supporting beams; hand pumps working night and day to get rid of the ever invading seepage. Wet bunks, wet blankets, wet uniforms; officers and men blear-eyed from the smoke of lamps, pasty-faced from lack of sunshine and fresh-air, sticking to it, "carrying on" in the best tradition of their wonderful race! Why did men do this? It was inspiring, mystifying, and incomprehensible - all at the same time! What was it: spirit of adventure, juggling with the unknown, gambling instinct, blessed release from the drab monotony of everyday life - or that intangible something we call Patriotism? "Quien sabe?".

Kemmel Hill

Kemmel hill we did on foot: it really is no more than a large wart on the landscape, but in flat Flanders, a few feet elevation gave a point of vantage and constituted an important observation post. The ruins of an old tower gave mute evidence of it's former importance; early in XIVth century, the French had defeated the Flemings on the plains between Cassel and Kemmel; English, Spanish, Teutonic, and French armies fought many a battle on these same grounds; small wonder is it that Belgium is known as the cockpit of Europe!

During the World War, Kemmel hill was overrun by the Germans who, thought sheer weight of numbers and superiority of artillery, compelled the brave Belgian army to abandon all of the fatherland except a small corner; it was conquered by the British and - in August 1918 - retaken by the Germans when held by a French army corps. In January 1918, it lay behind the British lines and was not occupied by troops. Colonel G... had just explained to me that the hill was deserted when I spied thin curls of smoke issuing from two or three of the many dug-outs which, like rabbit warrens, dotted the hill-side. We both saw the telltale evidence of human life at the same time! I looked at the Colonel; he looked at me in confusion, cleared his throat once or twice, hesitated an instant longer, and then - confidentially, oh! very confidentially, you know, old chap - informed me that there were always a number of Australians who, in periods of calm, went A.W.O.L (absent without leave) and lived a care-free life in these and other dug-outs back of the lines, devoid of hum-drum guard or patrol duty. Said the dear Colonel to me: "When a show is on, these chaps are tipped off by their comrades and all show up just before going over the top. We can't shoot our best shock troops, you know, so we wink at the A.W.O.L. Jolly practical solution, say I, what?"

Inspecting the First Aid Stations

In order to inspect the first-aid stations, I was "passed on" from H.Q. to Bramthock where Colonel O'G, D.D.M.S of the 33rd British division, was stationed. A wooden hut was assigned to me for sleeping quarters. It was surrounded by tiers of sand-bags four feet high with a narrow gap in front just wide enough to let one slip through the door. Overhead, there was no protection but the roof; a direct hit was "just too bad"! Inside the hut, an improvised table, ditto stool, ditto cot. The latter consisted of four rough boards as a frame, over which chicken wire had been stretched and nailed down. No mattrass or pillow; two blankets to lie on, two to roll up in. The exact size of the wire meshings was neatly stamped on my back and sides for many days! Surrounding the cot was another wall of sandbags, completely encircling it, so that going to bed somewhat suggested jumping into a high, old-fashioned bath-tub! Heavy blankets hung over window and door, so that no ray of light could be seen from outside and act as a mark for enemy night flyers. Removal of uniform and addition of an extra pair of woolen socks completed the bed-time toilet; life was too uncertain in these parts for pyjamas, and sleep without heavy Jaeger underwear was out of the question. Gas-Mask, helmet, and flash-light lay on the table, ready for instant use. Yet, in the midst of incessant warfare, the British gentleman, true to tradition, clung to many of his life-long habits. Shoes, puttees, and Sam Browne belts were daily polished by his batman. Early each morning, hot water for shaving and a small collapsible canvass tub for sectional bathing was brought in with the classical cup of tea without which no self-respecting Britisher can begin the day! Dressing for mess consisted of changing to fresh shoes and puttees and removing one's Sam Browne belt. Afternoon tea was never omitted!

"How did you get on with the British?" was a question often asked me during and after the war. Mon Dieu!, as the French would say, how does one gentleman get on with other gentlemen - by all remaining gentlemen, of course! Quite simple, if one holds steadfastly to one's instincts, the latter have a background of gentle breeding and the Golden Rule is adhered to. The American "mucker" or British "Tout" were unfortunate exceptions. Tweaking the lion's tail has long been a favorite pastime of yellow press and upstart; even the worm will turn, why not the British lion? Fundamentally, the average Englishman is shy, tries desperately to hide this trait under a cloak of assumed or sub-conscious aloofness, and is, often, deathly afraid of being "spoofed" by "smart" Americans. No matter how different customs and habits are all over the world, "a gentleman's a gentleman for that".

"Our best first-aid stations in this sector are the 'pill-boxes' we took over from Fritz when this area got too warm for him; you had best begin with them" said the D.D.M.S. to me at mess. A captain R.A.M.C. (Royal Army Medical Corps) was assigned as my guide, and five in the morning decided upon as the safest hour to venture forth because, visibility being nil, Fritz could not possibly observe us. The plan was to inspect the aid-posts, give me ample time to make notes on the treatment and evacuation of wounded, and return to H.Q. before the enemy laid down his daily barrage of shrapnel. Something went wrong, somewhere. I could not then and cannot now, in retrospect, flatter myself that the Flanders baptism of fire to which I was subjected was entirely put on in my honor! But let's not anticipate. We were driven as far forward as the main road allowed. When the latter disappeared in a veritable sea of shell-holes, we continued, on foot and in single file, along the duck-board path which constituted the only means of ingress and egress to the front lines. These, in many places, consisted of shell-holes, trenches being impossible to construct or maintain, on account of the mud and water. Each soldier on front-line duty was supplied with a tarpaulin, which he laid over the bottom of his particular shell-hole, after bailing out the water as best he could. The poor fellow could not stand up without exposing himself to machine-gun fire; he could not lie down; he squatted or sat. Whenever possible, these soldiers were relieved every forty-eight hours. A certain unusually exposed section of the Ypres salient had not been found economical, [and] the C.O. [was] able to prove that the casualties incident, to the relief of troops, through barrage fire was greater than those caused by trench foot.. The soldiers in this particular sector were consequently required to remain at their posts for five-day periods. Trench foot regulations, though admirably thought out and theoretically excellent, could not be observed in many cases. One typical British Tommy whom I quizzed in the C.C.S., tersely expressed himself as follows: "If oi gets me feet out of the bloody mud and water, oi slips in an drowns: arfter twelve hours of duty me feet are so bloomin swollen that oi carnt tyke hoff me shoes: ow in ell can oi 'ave me feet rubbed or chynge me bloody socks, oi arsks you?" Gunny sacks, loosely tied around the legs, instead of puttees, and extra large shoes were supplied [to] these men: rubber boots were a failure, because their interior soon became soaked with perspiration, socks became damp, and the men developped chillblains.

Just what is, or was, trenchfoot? Using simple terms familiar to the layman, trench foot was the result of prolonged exposure to cold and moisture, standing still, and interference with the blood circulation as socks, shoes, and puttees contracted from dampness. It begins as a sensation of increasing coldness of the feet, a feeling of "dead feet", followed by severe "toothachy" pains and swelling of toes, heels, and ankles. After a few days, the skin cracks, toes and heels bleed, and infection frequently occurs. The pain is worse at night, but the swelling disappears [only] to return again after wearing socks and shoes a few hours. It was my misfortune to arrive in the British war zone with an almost healed case of trench foot. The trouble promptly recurred and daily, from noon on, I suffered the tortures of the damned! As heat increased the pain, I hit upon the expedient of exposing my feet to the cold night air and managed to get some sleep in this manner. It was easy to slip on shoes and socks in the early morning hours when feet were half frozen by the cold, but hard and painful to pull them off when dressing for mess!

Situated about a thousand yards from the interconnected shell-holes which formed the front lines was our destination, a "pill-box". When the Germans held that sector, they had built ferro-concrete machine-gun forts or "strong points". Each miniature fort was approximately twenty foot square and eight foot high in the interior chamber. The thickest walls had, naturally, faced the old British lines; to remedy this weakness, the English Engineer corps had reinforced the thinner walls, which now faced the Germans, by means of heavy rails or tiers of sand bags. As I entered the pill-box, who should grasp my hand but one of the young medical officers I had crossed over with on the S.S. St Louis, in September 1917! He had been assigned to temporary duty with the British army and was unflinchingly doing his duty, far from friends, still further from the Paymaster, destined to form part of the "forgotten and unsung" legion of American officers who, in the stress of war, detached from their units, failed of well-deserved promotion, and returned to civil life with a sense of injustice and bitterness in their hearts. My young fellow-countryman was the only officer in the aid-station; he had under his command a sargeant, two orderlies, and eight squads of litter-bearers. All slept in the most sheltered part of the cement building; a wooden lean-to was used as a kitchen. On one of my visits, one of these kitchens was completely destroyed by a direct hit while another shell fell and exploded immediately in front of the pill-box entrance - I happened to be walking on the duck-board path a bare two hundred yards away! The food question was very practically solved; water was brought up in five-gallon gazoline cans from the nearest filtered and chlorinated water-supply station. Hot food also came in the same sized containers, placed in gunny-sacks, and surrounded by a three inch layer of hay. The food remained hot for six hours but could be warmed up in the kitchen by means of a "Tommy's cooker", a small lamp containing solidified paraffin as fuel. Needless to say, everything had to be carried on the backs of soldiers: no vehicle or horse could possibly use the light duck-board paths.

By nine in the morning, I had completed my observations and jotted down whatever seemed to have value from a surgical view-point. A solemn hand-clasp, a mutual "cheerio", and the young U.S. Lieutenant had vanished into the gloomy darkness of his surgical fort while the captain R.A.M.C. and I retraced our steps along the duck-board. As the direction of these paths changed every twenty yards, in order to avoid enfilading fire and wholesale destruction, one almost became dizzy, especially if one's thoughts became disturbed by the scenes of former carnage which dotted the landscape. Here, a broken artillery caisson with dead horses still in harness; there, helmets, both German and English, half sticking out of the mud; an arm, a leg, or torso partially submerged in the murky, greenish-yellow water which filled old shell-holes; nauseous fumes rising from the surface of what had once, and is now again, a fertile land largely devoted to the raising of hops, only the intense excitement of war prevented one from succumbing to the hideousness of one's immediate surroundings.

Barrage from Fritz

Hardly had we gone five hundred yards from the pill-box when, without a second's warning, a perfect barrage fire of shrapnel fell in a straight line in front of our erstwhile refuge. At intervals of one minute, the infernal shells fell ten or twenty yards closer to us - a perfect demonstration of a creeping or advancing barrage. To attempt to run back through it, in order to reach the pill-box, would have been suicide; only one direction, and one duck-board path lay open to us; we continued on our way, rapidly loosing that dignity and military bearing which authority breeds in an army! "Awfully sorry sir!" said my guide "I assure you Fritz has never before started firing at this early hour of the day; we planned the whole trip at H.Q. so that your investigation tour would end before the daily show began; carn't understand what the Kaiser is about, but I'm afraid we shall have to do the double-quick sir." A glance back showed us that the barrage was now a bare hundred yards behind us and making longer jumps forward, so double-quick we did. At a point where two duck-board paths converged into ours, detachments of a labor battailon which had been doing repair-work came running full speed, passed us without the formality of recognition, and sped on, hurling sanguinary epithets at their invisible enemy. Closing the file of sprinters was an old sargeant who, as he ran, half turned towards the captain, and shouted: "Run like 'ell, sir! they allus get's this blinkin corner; hit's somethink hawfull!" Run like hell we did; Fritz got the corner all right, just as my friend the unknown sargeant said he would - but the Captain and I had left the spot a few seconds before! The very fascination of the scene, the impromptu race with death, clever manner in which our retreat to the pill-box had been cut off by the first salvo, and admiration for the mathematical accuracy of fire-control, causing straight waves of mud and water to resemble an advancing ocean tide, prevented one from getting really scared, but I confess that my knees wabbled a little, and that my heart went rapidly pit-a-pat for a few minutes! It also gave one an excellent idea of the wastage of shells; hundreds of them were fired in a single barrage lasting only a few minutes; the labor battailon detachment was interrupted in it's work and had an additional section or two of duck-board path to repair, the Captain and I finished our tour a little ahead of scheduled time but - nobody was hurt! Such incidents explain the peculiar war philosophy which one quickly acquires at the front. One may be killed or wounded or gassed at any moment of the day or night, but the chances are so immeasurably in your favor that, aside from taking every possible precaution, one pays little or no attention to the noises around one. "That corner" marked the end of the creeping barrage, but naughty Fritz would not give us any peace, so he started elevating his range and exploding the shrapnel in the air above us. The scream and whiz of shells was most annoying at first, because we craned our necks to watch the explosions and narrowly escaped falling off the duck-board into the mud and water surrounding us. We were soon able to roughly estimate the general direction of the shell's trajectories and probable time of explosion; so far, all of them had exploded on our left and several hundred yards up in the air. One shell however evidently had a faulty time-screw: we heard the beastly thing directly behind us, whizzing crescendo then diminuendo down, down, down towards us, felt certain that it was going to make a direct hit on one of us and - laugh if you will, but it's a fact nevertheless - both of us instinctively bent down in somewhat leap-frog position and awaited the dread instant. It fell, miraculously, in the center of a shell-hole a bare fifteen or twenty yards off, exploded and covered us from head to foot with - mud and water, whew! Had it been a high-explosive shell instead of shrapnel, a fragment of casing would certainly have hit us. The Captain and I shook hands silently, wiped our faces as clean as we could and, to show our utter indifference to Fritz's impolite attempt at annihilation, nonchalantly lighted cigarettes, and strolled back to where our staff car awaited us for the return journey!

Just as we started back towards H.Q., we heard a soldier shouting: "Blyme, Bill, the sausage is off!" Looking on our right, we witnessed a most exciting scene. One of the British stationary observation balloons had suddenly snapped it's cable. Up it shot in the air, the wind rapidly driving it towards the enemy lines. With the help of field glasses, we were able to see the observer climb out of the basket with the parachute strapped to his shoulders and jump into space; the parachute gracefully unfolded and let him slowly float down towards the ground. We dashed to the nearest post for information. A shout of joy went up a few minutes later when the telephone rang to announce the English soldier's safe descent behind his own lines and on a dry hillock! Monotony of war? Bah! There's something exciting - comic, tragic, amusing, or sad - going on all the time, if you are lucky enough to get assigned to the zone of activities!

Responsibilities of a Military Surgeon

At H.Q., the officers tried to make a hero out of me, and drinks were somewhat lavishly passed around with the obvious intention of calming my supposedly shattered nerves with the flowing cup! I finally had to tell my good colleagues that I had already made numerous trips along the French lines and been in a number of rather "warm" spots. By mess-time, my poor feet - quite forgotten during the morning's exciting events - began to swell up and throb quite painfully. Experience had taught me that a change to extra-size socks and shoes, plus exercise, gave the most relief so, in the afternoon, I suggested walking over to the Advanced Dressing Station which corresponds to our Divisional Field Hospitals. It had been installed in the ruins of an old chateau, iron rails and corrugated iron sheeting replacing the destroyed roof. Four layers of sand-bags surrounded what remained of the walls and double sliding doors gave good protection against gas attacks. A number of gassed soldiers were being treated when I arrived, pending their removal to the rear. On account of the frequency of bombardments, it was only relatively safe to transport the wounded from one shelter to another, either at night or in the early, foggy morning hours. Inhalations of steam, with tincture of Benzoin added, was the accepted treatment at that time. There being only one nozzle, the slightly gassed soldier slowly walked around it, each man taking turns at inhaling the soothing vapor. The severe or "lying" gas cases had the nozzle brought to them, from stretcher to stretcher. The slightly gassed men envied the stretcher unfortunates;

the latter would eventually land in beloved "Blighty" and probably never be used again as field troops, whereas the former would merely be sent to the Corps Rest Station for recuperation and return to duty with their old outfit. Perhaps the greatest responsibility a military surgeon has to face during war lies in the accurate sorting out or "triage", as the French called the process, into casualties able to return to duty in a few days, those needing treatment in Evacuation hospitals but still able to be salvaged, and those whose fighting or even military days are over. Malingerers will turn up in any army; race, religion, heredity, and previous environment, all have to be considered and given their proper weight in making such a decision. Consultants and medical boards of review were frequently appealed to for final decision in complex cases.

Corps Main Dressing Station

From the advanced dressing station, a narrow gage railway brought the wounded to the Corps Main Dressing station. Evacuation was by means of two specially fitted up flat cars with a carrying capacity of sixteen stretchers and eight sitting patients. The trucks were drawn by a small gazoline engine, the round trip of about eight miles taking an hour. Motor ambulances, although subjected to shell-fire, also reached the advanced dressing station, especially at night-time.

In order to visit the Corps Main Dressing station, I was again "passed on" to yet another D.D.M.S. and installed in the cellar of a demolished house just around the corner of the square in which stood the famous cathedral and "Cloth Hall" of Ypres. My roof was a criss-cross jumble of railroad ties, sand-bags, and corrugated iron sheeting: water trickled down the walls and dripped on my bed, and rats played games over my body as I slept! Bombardments were of the incessant variety; shells were either whizzing over the martyred city or falling on the ruins of the Belgian army barracks and store houses. Even on a Sunday, a shell made a direct hit on a large Y.M.C.A. tent in the midst of service, killing and maiming a large number of soldiers. It was noticeable that, when walking around the city of Ypres, officers and men instinctively sheltered themselves behind ruined walls and double-quicked when crossing a street. Life was indeed a gamble in that salient, with "Lady Luck" holding the stakes!

The spacious cellars of the Court House was used as the Corps Main Dressing Station, the evacuation of wounded from two sectors converging at this point. It was the first place where emergency and life-saving operations could be performed amid proper surroundings and under bomb-proof shelter. While the building itself was a mass of ruins, the immensely thick arched cellar walls had resisted every bombardment. Electrically lighted, well heated, and clean, the series of connected cellars had space enough to contain fifty stretchers at one time besides six operating tables with their necessary equipment. The knowledge that no shell, however powerful, could penetrate the mass of stone and brick which lay above one's head was conducive of a curious sense of peace and aloofness from the carnage and turmoil overhead. Surgeons and orderlies lived near-by in equally bomb-proof cellars, were well fed, and could usually enjoy fairly regular hours of duty and off-duty. As one passed from front to rear in any army zone, the atmosphere around one, the psychology of life, underwent perceptible changes. Thousands and thousands of our officers and men in France never heard the sound of cannon, ta-rac-tac-tac of machine guns, explosions of trench mortars or hand grenades; stationary suasages, "dog-fights", and gas fumes were unknown to them as they toiled at the Bases or Lines of Communication, carrying out duties fully as important and essential as those of combat units, but devoid of physical danger. Not being able to visualize battle conditions, experiences of comrades from the front made little or no impression on them. This probably explains the oft-repeated complaint at home that officers and soldiers who had actually seen the war were the ones least inclined to discuss their experiences. We felt - and still feel - the utter futility of trying to paint a picture which could not possibly evoke any sign of recognition or appreciation from our audiance; they had no yard-stick of previously stored-up memories by which to visualize what we were talking about!

A Stroll in Ypres

Strolling around Ypres - hugging walls and skipping rapidly across streets, you may be sure -, late one afternoon when not on tour of inspection, the melancholy sounds of a bag-pipe came to me from the deep cellars of a destroyed mercantile house. Headed by the piper, a company of Scottish infantry marched up the broken stairs, assembled in the street and started towards the front lines to relieve another outfit in the shell-holes and muddy waters. Civilians are wont to picture soldiers starting to battle as gay, hilarious, bedecked with flowers, and eager to be "up and at 'em"! The demeanor of officers and men, slow march, and sad "wail o' the pipes" were more like the beginning of a funeral procession. Not that the Scotties were less brave than any of the other contingents which made up the British army, far from it. The serious faces, heavy deliberate marching step, and woe-begone bag-pipe notes seemed to symbolize the murky, foggy hill-slopes whence these men came from. It was the "dour" Scotchman's temperament which craved a minor key in his music:

the doleful bag-pipes notes reminded him of heather and brake and Mavourneen handing him a dish o'grits. Perhaps this unchangeable difference in traditions and customs partly explains why so many of the British regiments remained all Welsh, all Irish, or all Scotch through the years. There is none of this pride of locality or race observable in the American, French, or Italian armies, although the "Blue Devils" of France and "Alpini" of Italy are selected for their native ability as mountain climbers.

In spite of the erratic German shelling which swept the Ypres salient, day in and day out, British officers could be seen enjoying their afternoon canter wherever they could find passable roads which were not shell-swept. Trotting or cantering nonchalantly, just as though they were in Hyde Park or trying out their superb mounts along some English country lane, the psychological effect of such calm and unaffected adherence to hereditary custom on their soldiers must have been excellent. No real attack was contemplated or expected; annoy the enemy as much as he is annoying you, keep a vigilant eye open against surprises - but don't forget afternoon tea and cake! That seemed to sum up the stolid British philosophy, the state of mind of a race that would carry on to the end with it's back to the wall, never despairing, never panicky or hysterical - always British!

The Auctioneering of Monsieur de Garenne

Late Months of 1917

Being a citizen of noble France, his real name was Monsieur Lapin de Garenne, but we crudely and rudely refer to his American prototype as simple Jack Rabbit!

The auctioneering of Monsieur de Garenne.
By
George Skitarnow

Prologue:

Of valiant deeds, involving mass attacks, heroic defense, or superb counter-offensive, much has been written. War correspondents, military "Communiqués", and post-war critics have familiarized the people with military events seen "en-masse"; few have stooped to recounting the minute, individual, tragic, or comic aspects of the World War. "The...nd Division stormed Montfaucon, or the...st Regiment captured Suippes", is all very well as far as it goes, but what Private Bill Jones or Corporal John Smith accomplished individually, rarely gets beyond the unit commander's final report. It is of the smaller details of the war that I would write, of the incidents which enlivened the horror, suffering, and monotony of warfare, of the smaller and therefore forgotten actors in the vast drama. The episode which I am about to relate actually occured in the battle sector held by the French 1Xth army to which I was temporarily attached as Chief Surgeon of an Evacuation Hospital in the late months of 1917, before our own troops were ready to "do their bit".

George Skitarnow.

Being a citizen of noble France, his real name was Monsieur Lapin de Garenne, but we crudely and rudely refer to his American prototype as simple Jack Rabbit!

His ancestral chateau, boasting of many entrances, exits, shelters, listening posts, moats, and dungeons - or, as the nobleman would have described them in correct military terminology, entrées, sorties, abris, postes d'écoute, fossés, and oubliettes - lay in the midst of a small but picturesque forest which we can, for the purpose of this story, place in - not on - the hill of Montrécourt, in one of the invaded sectors of this war-ridden land.

Montrécourt, as it's name plainly indicates, is not a long hill, and we would never condescend to call it a mountain - but they manage such things much better in France!

During the early months of the war, when dashing "Poilus" marched up the hill and then marched down again - until the General Staff told them that it was time to halt and beat back the invader, all of which we know they did, Monsieur de Garenne with his numerous family and retainers was mobilized (which means to be put on the move, in military parlance) and during this otherwise necessary and patriotic sequence of events, many of the more daring and therefore less cautious members of his family were sacrificed to the voracious appetites of the fighters! If you don't believe me, ask the next Poilu whom you meet whether or not he likes "Civet de lapin au vin" for a change! Stewed rabbit, as prepared by French cooks, is one of the most delectable dishes imaginable. The luscious tender bits of meat, aromatic seasoning, and wine sauce make of it a truly epicurean dish. Prepared by the average American housewife or cook, it is an insipid, tasteless horror!

As the war of movement ceased and trench warfare took it's place, the hill of Montrécourt became the scene of much terrific but useless and inconclusive fighting, both armies finally settling down along either slope, leaving a crest of hill a mile long and from thirty to three hundred yards wide for the "no-man's-land" of which civilians read so much about and knew so little. As this is the story of Monsieur de Garenne, I must pursue the narrative, leaving to future opportunities a description of this variegated territory. Suffice it to say that you may, at times, go duck hunting in it - if the ducks are flying and the bullets are not -; at other times, the mere sight of the crown of your ugly but efficient steel helmet above the trench top is sufficient to start a hurricane of bullets, which makes you duck down in a most undignified fashion! This makes three ducks in one sentence, but they are of different varieties or species, and the family is a large one......... as I was saying, Monsieur de Garenne, with a sadly reduced family, finally remained in sole possession of Montrécourt crest. He could not advance, and he could not retreat; his flanks ended in two impassable swamps on either side of his domain.

From his favorite "poste d'écoute", or listening post, he could sadly gaze upon the somewhat battered, but still extant cabbage patch in which, from the dawn of French "soupe-au-chou" or cabbage soup history, he and his had been wont to satisfy their nocturnal appetites! The luscious cabbage heads beckoned irresistibly; and one after the other, the de Garennes - legitimate descendents and bar-sinisters - attempted the impossible and improved the "rata" of the day! The "rata", O initiated reader, is the classic, historic French army ration, a wonderful concoction demi-soup, demi-stew in which the "cuistot", or army cook, has amalgamated all of the daily supplies with mysterious seasonings and "arrosage-au-vin", or wine sauce, so that, as you approach the company kitchen after floundering through miles of trenches, it's aroma fairly dilates your nostrils; and, a severe, imperative, insistent, and not-to-be-denied pang seizes you in your tummy as you smilingly accept not a demi-bowl, but a demi-platter full!

The de Garenne days, many at first but slowly becoming more and more infrequent, were invariably red-letter days, the "cuistot" being rewarded for his culinary ability with a package of cigarettes and an extra pint of "Pinard". Pinard is wine in poilu slang; nobody knows when, where, or how the name came into existence, but wine is no longer wine in the French army; it's simply pinard, just as a French soldier is always affectionately referred to as a "Poilu" or hairy one. This last nickname, however, has an authentic pedigree. During the first weeks of August 1914, the rapid marches and counter-marches of the French armies made it well nigh impossible for officers or men to shave, even at irregular intervals. Beard raising became universal and, as wounded were evacuated to the rear, they naturally caused the nurses to lovingly dub them "mon pauvre poilu" - my poor hairy one. Poilus they remained to the end of the war!

Finally, as we must come to the important part of the story, Madame Mère de Garenne attempted a desperate sortie or sally; because, having always been accustomed to live in a certain style, she did not see why a few wire entanglements, a trench or two, and cautious soldiers sheltered therein should deprive her of her heart's - no, her stomach's desire! Her family had melted away, her retainers had departed without giving notice, and Madame felt that intense desire to whoop it up, which the French so aptly call a "crise-de-nerfs" or hysterical attack. In no uncertain terms, she lectured her spouse on his inactivity in her behalf, asked him why he didn't do something to end this suspense, laughed, cried and shrieked in bunny fashion, and ended the scene by telling Monsieur de Montrécourt that he was an old stick-in-the-mud anyway, that she was sick of the whole thing and was going to eat cabbage or die!.

She did...... the cuistot received two packages of cigarettes and a quart of pinard for his share of the entertainment, and everybody felt very happy, that is, everybody except Monsieur who now remained, still the noble cavalier "sans peur et sans reproche" but also, alas! sans wife and sans children!

Thus began for Monsieur de Garenne, last of his race in Montrécourt, a life of constant vigilance and unending solitude. He submitted, as a true French habitant should and did, to his war rations. The tenacity of plant life, in the face of soil upturnings, explosions, powder-smoke, and poisonous gases was one of the marvels of the war. Even in hell-ridden sectors such as the Chemin-des-Dames or Verdun, nature fought the most fiendish instruments of warfare yet devised by man and won out - in patches! Each bombardment of Montrécourt crest destroyed some new patch of verdure or uprooted a tree or two; mustard gas made it impossible to visit certain favorite haunts for days on end; nevertheless, de Garenne "carried on" in the best traditions of his cotton-tailed race. His breakfast, or "petit café", consisted of plain grass which he called "fines herbes" - principally parsley and chervil - in loving memory of the kitchen garden he used to visit in the good old peaceful ante-bellum days when he only had to dodge Médor the watch dog and the village poacher, or "braconnier", in order to reach the cook's paradise and help himself to his heart's content. The reader must not gather the false notion that our hero ventured forth to his meals in the daytime. The sad fate of his entire family had impressed upon him the utter folly of such a procedure. True, he had in bygone days ventured forth for an extra juicy nibble or two at an hour when he should by rights have been slumbering in his deepest hole, but not now! What to de Garenne was breakfast actually corresponded to the twilight meal of a night-watchman; he partook of luncheon somewhere around midnight, and supper was daintily sucked just as the dreaded "zero-hour" of battle cast it's terrifying shadows across the land!

For lunch, he nibbled a square of ash-tree bark, a not very satisfying repast, and in the misty dawn, having omitted all main courses - which he called "plats de résistance" - because none were available, he permitted himself the luxury of a "sève de tilleul" - sève or sap being the equivalent of tea or tisane in the de Garenne milieu or circle. Be it known that, in France especially, an infusion or decoction obtained by pouring boiling water over the dried leaves of the linden or lime tree, is a favorite evening drink supposed to calm one's nerves and facilitate a good night's sleep. Families who do not follow this secular custom simply do not belong! Thus does true aristocracy maintain and assert itself under the most adverse circumstances! You cannot belong to high life if you omit the cup of linden tea before retiring. It is hot, it is yellow, it is of course very much sweetened and, above all, it is quite the thing to do! Noblesse oblige!, therefore Monsieur de Garenne, snapping his hind feet on the ground with a gesture of supreme distinction, continued nightly to dull the plebeian sense of hunger which occasionally threatened to master him, with linden-tree sap, until the few trees which his domain boasted dried up and refused to do their duty.

"Oh rats! oh sacred name of a pipe! this must come to an end, "j'en ai plein le dos" said de Garenne to himself when his last linden-tree had failed to yield the faintest drop of sap. Under similar circumstances, his English cousin would simply have said "I'm fed up" while our less distinguished, but more vigorous jack-rabbit would have opined "enough for mine"! The situation was truly a desperate one. Hopping into his longest and deepest burrow, he tried in vain to get some sleep. Slow starvation or sudden death, no other alternatives seemed open to him but, with the blood of a long line of noble ancestors coursing in his veins, he was the last one to give up the fight for freedom and free meals. Cogitating with his whiskers and tapping the ground with his stubby tail to emphasize a favorable point or dismiss a foolish one, he finally dozed off. As soon as the last ray of light had penetrated his bed-chamber, he awoke, kicked his aristocratic heels together, and ventured forth with a fairly definite plan of campaign in his mind. Under cover of darkness, Monsieur de Garenne made a cautious but thorough survey of what he ironically termed his defenses. At a point where no man's land was a bare thirty yards wide, he spied some German soldiers clearing away their own entanglements, preparatory to an attack. Such preliminaries were familiar ones to this war veteran. From time to time, it becomes quite necessary to know just what type of enemy is facing you, which Division that particular company or bataillon belongs to. Certain units are known to be specially trained as shock troops or are commanded by officers whose personality and knowledge of military science has inculcated a superlative degree of esprit-de-corps to the troops under them. Their secret shifting from one sector to another under cover of darkness is interpreted as a sign of near-future activity elsewhere; the arrival of less vigorous units, of reserves, means that that particular sector is to be a calm one for the time being. All of this information is obtained through the capture of a few prisoners in the course of a local, purely informative surprise attack. "Voila!" said de Garenne to himself as, having nibbled the last square of ash-tree bark, he retraced his steps towards his ancestral holes, "I shall sleep until early dawn and then escape into enemy territory before daylight appears. The Germans have so far not shown any disposition to disturb my race, and I can easily find some of my old ante-bellum haunts where I will be allowed to live in peace and plenty". He hopped into one of his oubliettes and settled down for the balance of the night.

Now Alphonse the Poilu, first-class soldier of France, had also noticed the German activity and notified his sargeant and companions of the impending sortie.

It so happened, fortunately for the story and the "rata" of the day, but most unfortunately for the last scion of the noble house of Montrécourt that his appearance opposite the German gap, in the early dawn of the following day coincided to the second with the beginning of the attack. The enemy suddenly, and as silently, removed a narrow strip of barbed-wire entanglements which they had previously cut with special steel shears, leaving an open pathway leading towards the French defenses. As the heavy helmeted "Feldgrau" soldiers of the Kaiser, wire shears and grenades in hand, swarmed over the trench top and plunged across no-man's-land, de Garenne started his headlong dash for liberty and cabbage. A terrific machine-gun fire stopped both at the same time! Our unfortunate hero uttered a single agonizing rabbit squeal, tapped the ground convulsively with his hind legs.. once.. twice.. and then lay extended in the center of the cleared space. The noble race of Montrécourt was no more!

The surprise element - so essential to success in an informative night raid - having utterly failed, no further attempt was made to penetrate the French entanglements. Cautiously pulling their battle casualties back into their own trenches and evacuating them to their bataillon aid-station, the enemy prepared to defend the gap. Repairing barbed-wire entanglements by daylight is suicidal and a none too healthy occupation even under cover of darkness!

Alphonse and his comrades were eager to still further punish the attempted invasion of their trenches, but all calmed down under their sargeant's cautious reprimands. The peace of watchful expectancy again permeated the hill-crest of Montrécourt. Monsieur de Garenne remained on the field of honor, in full view of Franz and Alphonse. The latter volunteered to make a sortie and obtain the coveted prize - the company cook would still have time to perform the necessary culinary rites and regale his comrades with the now all-too-rare rabbit stew with wine sauce. All were hungry; breakfast, in view of the impending raid, had been out of the question. The brave Poilu argued with, entreated, cajoled his sargeant - all in vain -; France needed her first-class soldiers for sterner purposes than the mere salvaging of a dead jack-rabbit! Ah! woe is me! Name of a Pipe! Sacred name of a name! muttered Alphonse, moistening his lips as he voraciously gazed at the noble remains of Montrécourt. "Comment faire? Comment faire? What to do? what to do?"

"Why not ask the German's permission to pick up the rabbit?" suggested the sargeant who always invited himself to rabbit-stew feasts.

Up went a white flag!

"Ohé! Franz! how about the rabbit?"

"Mein!" shouted the Prussian, holding his own white flag high up and popping his head cautiously over the trench top. "What! Sacred blue!, a French rabbit, lying on French soil, killed by French bullets - and you claim it's yours! Quel toupet! What impudence!"

"Mein" growled the German, threateningly. "Come and get it then" said Alphonse with a grin, placing his machine-gun in position. Franz did likewise, and the two warriors realized that negociations were not taking a very satisfactory turn.

"Loud sneered the doughty man of France,
Take it if you dare, Ho! ho!, Ha! ha!
The French for "pish" said soldier Franz;
Said Alphonse," the German for "bah!"

"The French for "pooh", our Deutscher cried,
The German for "pig" said A.. in French
And so, with undiminished pride
Each remained in his own trench!"

Thus would W.S. Gilbert, of Gilbert and Sullivan fame, have attempted to describe the dead-lock which threatened to be prolonged indefinitely.

"Ich kaufe! I buy!" shouted a German sargeant after a few minutes' hesitation, holding out a package of tobacco.

"Moi aussi! Me also!" retorted Alphonse, producing two packages of cigarettes!

Temporarily forgotten was the war; in the tense excitement which an auction invariably brings forth, caution was thrown to the winds, heads popped out over the trench tops in utter disregard of instructions, and voices were raised higher and louder in order to overcome the noise of artillery fire. No patrol passed by, or a summary courts-martial and demotion of sargeants would have resulted. The offers grew larger and larger, neither side being able to produce sufficient quantities of tobacco or cigarettes to close the bidding.

"We cannot loose such a good opportunity" said the French sargeant to himself as he realised the probability of a stalemate in the procedings. Producing a new box of American cigars which his "war god-mother" had given him and which he had been preciously saving up for his next furlough, he boldly climbed out of the trench and faced the enemy, unarmed. "Cigarres Américains" he shouted, brandishing the box over his head!

"Wúnderschön! Kolossal! cried overjoyed Franz who had often sampled the brand when he was kellner at the.......in Chicago.

France won - as she deserved to win. Alphonse and Franz simultaneously marched to the center of no-man's-land, the former laden with the irresistible perfectos.

The box was handed over, Monsieur de Garenne tenderly lifted up by his ears, and the two satisfied warriors regained their respective trenches!

Epilogue.

An interval of one hour has elapsed!

"Dûmmkopf Franzöz" growled Franz, blowing wonderful rings of smoke over the top of his trench. "Such tobacco!, such aroma!"

"C'est la revanche! it is revenge!" laughed Alphonse the hairy one, tossing the last portion of Monsieur de Garenne's bony framework under Franz's nose!

"C'est magnifique, mais ce n'est pas la guerre" muttered ye scribe, sighing for the copy of "Bab-Ballads" which he had failed to put in his regulation army locker!

A Wartime New Years Dinner

Tuesday, January 1, 1918

Officers of other units privately assured me that, in season, duck-hunting in these same marshes was so excellent that a sort of entente cordiale or mutual understanding grew up between opposing units and that both Germans and French refrained from firing at each other, providing they appeared with fowling pieces instead of army rifles!

A period of relative calm prevailed along the entire front. Gigantic military plans, of which we were, at the time totally ignorant, occupied the minds of Army Headquarters; for us of the Medical Corps, our duties were light, convoys of wounded easily taken care of, and gossip was correspondingly voluminous and fantastic. Spasmodic shelling of railroad junctions, stations, and military depots seemed to be on the increase, but we paid little attention to these daily occurrences. As a matter of fact, the Germans were methodically getting the range of our lines of communication in preparation for the March onslaught.

A splendid New-Year's supper was offered to the entire staff by the Directrice of the hospital. General Taufflieb and his A.D.C. Captain Froment-Meurice arrived for the occasion. Besides myself, the American contingent consisted of Major John J. Moorhead and Lieutenants Bolling, Stephens, and Schrock. Medecin-Chef Cassan, lieutenants Andrieu, Louvard, and Dallet, the latter being the "Gestionnaire", or Quartermaster, of the hospital, completed the party. The nurses had their own mess-hall and never mingled with the officers during meal-time, but were occasionally invited in for a short round of dancing to the dubious tunes of an antiquated Victrola. The chief topic of conversation was, naturally, the probably duration of the war. The French were convinced that the final campaign of liberation could not occur before 1919, when the American Expeditionary Forces would have attained their maximum strength and be fully trained. With the exception of those of us who were actually taking part in the war, few Americans realise, even to-day how much intensive training our troops had to go through after they reached France before they were judged capable of manning the front lines. Efficiency on the parade grounds and maneuvering in the face of an enemy are quite different problems. The creation of a General Staff was a gigantic problem in itself; commanding an isolated army post in the States was child's play compared to the coordination of a Division. The lack of trained Cadres worried the French more than anything else about our young army. The final casualty lists proved conclusively that we had to rely on four divisions for all of the important battles we took part in. The First, Second, Twenty-sixth, and Forty-second divisions constituted our shock troops and had to withstand almost all of the heavy fighting. This is not a criticism of our other divisions; the latter simply lacked the requisite amount of training. The Germans, with their marvelous spy system, knew these facts and wisely thought unsuccessfully tried to end the war in 1918! It would have been impossible to take offense at the kindly, purely constructive criticisms of our French "Associations"; they were, quite naturally anxious to use the American Expeditionary Forces to their full capacity but, at the same time, the French General Staff dared not run any military risks which might have courted disaster. After all, they were defending their own country, not ours. I never to this day can reconcile the use of the word associate instead of ally; it suggested an aloofness, a semi-independant and limited cooperation with the French, English, and Belgians, and failed to harmonize with our grandiloquent official declaration in April 1917 that the United States placed her last dollar and last man at the disposal of the Allies in order to win the war! We were all comrades in a common effort and should have remained close comrades during the sad aftermath of the war. That we did not remain good comrades after the Armistice could not have been foretold on that memorable New-Year's eve of 1918! Tongues wagged freely; topics both serious and gay alternated with the wines and liqueurs with which the banquet table was generously laden and, at the stroke of twelve, we clicked glasses, exsuded good wishes, individually and collectively, sang or hummed the Marseillaise, and hummed, only hummed, the Star Spangled Banner! Was there ever a group of Americans who could remember the words of our national anthem? As General Taufflieb and his A.D.C. made their farewells and departed, we heard shrieks of laughter outside and hastened to find the cause thereof. The nurses had discovered two antiquated sleds in the saddlery of the chateau and, there being a goodly coating of snow on the Gannelong hill, touching the gardens, they were merrily sliding down the straight but rough path which ended in a right-angle turn at the bottom. Grabbing our overseas caps and fur-lined gloves, we all joined in the fun. None of the French officers had ever managed a sled; as one after the other failed to make the turn and crashed against the wire fence, he would retire to nurse [a] sprained wrist or ankle or apply some antiseptic to a scraped nose or cheek! The Anglo-American contingent kept at it until the wee hours of the morning – the one and only sledding we had during the war!

A few hours of sleep and Lieutenant Schrock and I were on our way to General Taufflieb's headquarters at Ugny-le-Guay for a three day tour of inspection of trench warfare. The sector we were taken to bordered the swamps of the Oise river, south of the old city of Noyon. Unless frozen over, these swamps were quite impassable and offered ample protection against any surprise attack. Instead of the usual series of trenches and strong-points, small "fox holes" had been made, just deep enough to conceal a soldier sitting or kneeling. Six men to a mile of front line was deemed sufficient for defensive purposes. The guards had just been changed, and we were able to note the precision of detail in that area. Spying a loan tree near the margin of the swamp, the General asked the Lieutenant who was accompanying us if he had camouflaged a small trench mortar – called a crapouillot – behind that same tree. No, mon General; I didn't know there was supposed to be on there! But of a certainty, Lieutenant; come and I will show you. Sure enough, there was the emplacement for the mortar – but no mortar! See to it that this error is at once remedied. Oui, oui, mon General; I run at once to see to it! Even in trench warfare; even if there is no possibility of an attack from this direction, one must constantly warn the enemy that we are on the alert; a bomb from this emplacement can be neatly dropped in the German front-lines, beyond the marsh; it keeps our adversaries under cover and prevents them from observing our movements too closely.

Officers of other units privately assured me that, in season, duck-hunting in these same marshes was so excellent that a sort of "entente cordiale", or mutual understanding, grew up between opposing units, and that both Germans and French refrained from firing at each other, providing they appeared with fowling pieces instead of army rifles! As say the Italians "se non e vero, e ben trovato!" if not true, t'is well told! An Alsatian by birth, General Taufflied hated the Germans with a deadly hatred which almost passed all understanding. As a young "shavetail", or second lieutenant, just out of St Cyr, the French West Point, he had slipped into Germany, passed himself off as a German mason, and been employed in the construction of the series of forts which surrounded the city of Metz. That he had spies at his service, he proved by showing me a bundle of German newspapers which he received daily and read assiduously. Needless to say, he never revealed the method which enabled him to guage the morale of his enemies! Another Alsatian, General Hirshauer, whom I met in the Verdun sector, was, according to information, equally well supplied with enemy information. It was quite evident that the forty-seven years, which had elapsed since the Franco-Prussian war of 1870, had not dimmed the "amour de la patrie" in the minds of many, many Alsatians who daily risked their lives in order to help bring back their lands to France.

A la Rescousse les Poilus

February 1918

The higher command was necessarily French. This arrangement was, in the main, satisfactory, in spite of differences in language and customs. Whenever misunderstandings arose - and they were many - the basic underlying reason was inability on the part of officers and men to understand each other's language. Fortunate indeed were the Americans who could speak French and equally fortunate the French who could talk to us. No finer race exists than the French - when you know and can understand them.

A la rescousse, les poilus!

An amusing incident of the World's War wherein
French veterans
rescued
inexperienced American soldiers.

by George Skitarnow

Property of
George de Tarnowsky
Colonel Medical-Reserves
U.S. Army
30 N. Michigan Ave
Chicago, Ill.

When, in the early months of 1918, one of our future good shock-divisions was moved up to Sector "S" in northern France, it was judged best, for a variety of reasons which could not be discussed at the time but that we can now – in retrospect – lay down to training necessities, to intercalate our youngsters, company of company of Infantry and battery for battery of Artillery, with veteran French comrades or "Poilus" i.e. "hairy ones" affectionately so dubbed because they had had no time for shaving during the agonizing weeks of August 1914.

The higher command was necessarily French. This arrangement was, in the main, satisfactory, in spite of differences in language and customs. Whenever misunderstandings arose - and they were many - the basic underlying reason was inability on the part of officers and men to understand each other's language. Fortunate indeed were the Americans who could speak French, and equally fortunate the French who could talk to us. No finer race exists than the French - when you know and can understand them. It is a sad commentary on human intelligence to have to admit that the average person has a sub-conscious feeling of superiority when brought in contact with a fellow human being whose language he does not understand!

Here and there in our Division, a few "fire-eaters" grumbled at first because the fighting was too slow for their enthousiastic temperaments; it did seem hard at times to occupy a section of German trench, dispose of all of it's occupants via hand-grenade, bayonet, or "Kamarad" method - and then receive orders to retire back to one's old lines when one felt that you and your buddies could go on to Berlin and get there in time for lunch "Unten den Linden"! Had they, from the first, been able to visualize the labyrinth of second, third, and communicating trenches into which the enemy could pour in their reserves: had they known of the many "strong points" bristling with machine-gun nests from which a deadly cross-fire would have mowed them down before they could advance beyond the thinly manned first-line trenches, our impetuous youngsters might have felt less eager to attempt a "Promenade a Berlin"! Sector "S" being a purely defensive zone, i.e. one from which a penetrating attack en-masse could not be contemplated on account of lack of proper transportation facilities - railroads, canals, highways - without which at it's back no army would dare venture, the system of trenches and dug-outs was most elaborate. Seven feet deep and four feet wide, with a central gutter which, oftenunsuccessfully, tried to drain off rain water, the bottom of the trenches was overlaid with sections of wooden cross-bars, two yards long, built ladder fashion with cross-sticks two feet wide and two inches apart, which prevented the soldiers from slipping and floundering about in the mud during the rainy season. The English called these sections duck-boards, the French and Belgians "caillebottis". The direction of these trenches changed every fifteen to thirty yards, in order to avoid enfilading fire. Sand bags facing the enemy topped the trenches for added protection, with narrow intervals between bags through which the rifles or machine-guns could be thrust. At intervals of five hundred yards, more or less, a dug-out had been built by the French corps of Engineers; all our soldiers had to do was to protect the sector from surprise attacks; shelters and defenses were already there. Whenever possible, the dug-out was two stories deep, approximately ten and twenty-four feet below the bottom of the trench. It almost always had entrance and exit doors protected against gas attacks by heavy canvas curtains which, as soon as a gas attack was announced, was sprayed with a solution of baking-soda. The stairway, three feet square, was built at an angle of forty-five degrees. Wooden pillars, cross-beams, and corrugated iron sheeting for the roof, prevented the earth from falling in. The upper room was actually the guard house while bunks, arranged in two tiers, lined the walls of the lower chamber. Small wood-burning stoves gave out a reasonable amount of heat, and a ventilating pipe extended vertically from the lower room to one foot above the surface of the earth. A tin cap on a spring hinge enabled the ventilating shaft to be closed during gas attacks. Attached to the inner side of the cap was a long string, which passed down the air shaft into the lower room. At the first alarm, the soldier nearest the shaft had orders to pull the string taut and fasten it to a convenient nail. This simple device prevented the heavier-than-air gases from pouring down the shaft and choking the soldiers. When the gas attack had ended and antidotes had been sprayed around the upper end of the air-shaft, the string was loosened, and the spring hinge released the cap. As a rule, the air was fairly pure. When not on actual duty in the trenches, the soldiers led a fairly comfortable life, deprived of sunlight and means of exercise, but safe from gas attacks, gun and shell-fire. One or two "Primus stoves" in the upper room enabled the men to warm up their food, which was brought up to them at night. Excepting the detail on patrol call, most of the soldiers slept in the lower room, arms stacked in one corner, and trench boots off but otherwise fully dressed. Pyjamas would have been as incomprehensible as umbrellas! A period of front-line duty, varying with the locality and activity of the sector, was invariably followed by a turn in the "rest area", so that the monotony of trench warfare was reduced to a minimum. So much has been written about the horrors of military life that the many and successful methods of bringing comfort to the fighting men is often lost sight of.

A trench only represented an infinitesimal part of the gigantic battle-line and "detail operations" - as the French command called these local attacks - always had to be subservient to the general scheme of things. If Sam Amex was at times inclined to consider - very privately, you may be sure – French Headquarters as too cautious and methodical in it's estimate of the exact value of these detail operations, he soon learnt to look upon the Poilu, the "hairy one" as all French soldiers were affectionately dubbed, in the light of a big brother, big in wisdom, in his intimate knowledge of every nook and corner of the sector he occupied and, especially, in the art of out-guessing Fritz, over there on the other side of no-man's-land. Indian warfare, as depicted so fascinatingly in the exciting novels we used to devour in our happy school days, was simplicity itself when compared to the tricks which were constantly being elaborated by both groups of belligerants. Said a French general to me one day, in the course of a front-line inspection: "The real art of trench warfare does not consist in merely preventing your enemy from entering the trenches you are protecting: it lies rather in repeatedly springing some new surprise on him. If repeated too often, any system of attack will inevitably cause a satisfactory method of defense to be evolved. Constant novelty and constant surprise are the two great desiderata to be sought for. The enemy should never be allowed to feel secure and, as far as humanly possible, he ought to be induced to guess wrong. Each section commander should repeatedly say to himself: 'What can I do to the enemy that I have not done before?' He should regard the day as lost in which he has not been able to bring in some prisoners or inflict more losses than his section has suffered."

These tactics only applied to what was known as a calm sector - which simply meant that the thousand-and-one daily incidents with which the entire line bristled were never recorded in the official reports handed out to War correspondents. Such war books as "All quiet on the western front" and "Journey's end" deal specifically with just such minor phases of warfare. Many and many a time has a perfect hurricane of shells of all calibers been fired by one or both sides - either in foggy weather or, more especially at night-time - because a stray dog or venturesome jack-rabbit had appeared in no-man's-land and been taken for an enemy reconnaissance by some nervous sentinel! This probably sounds like gross exaggeration to the uninitiated, but it's the gospel truth! A night alarm spreads along the line with the rapidity of lightning, and noise actually seems to be contagious!

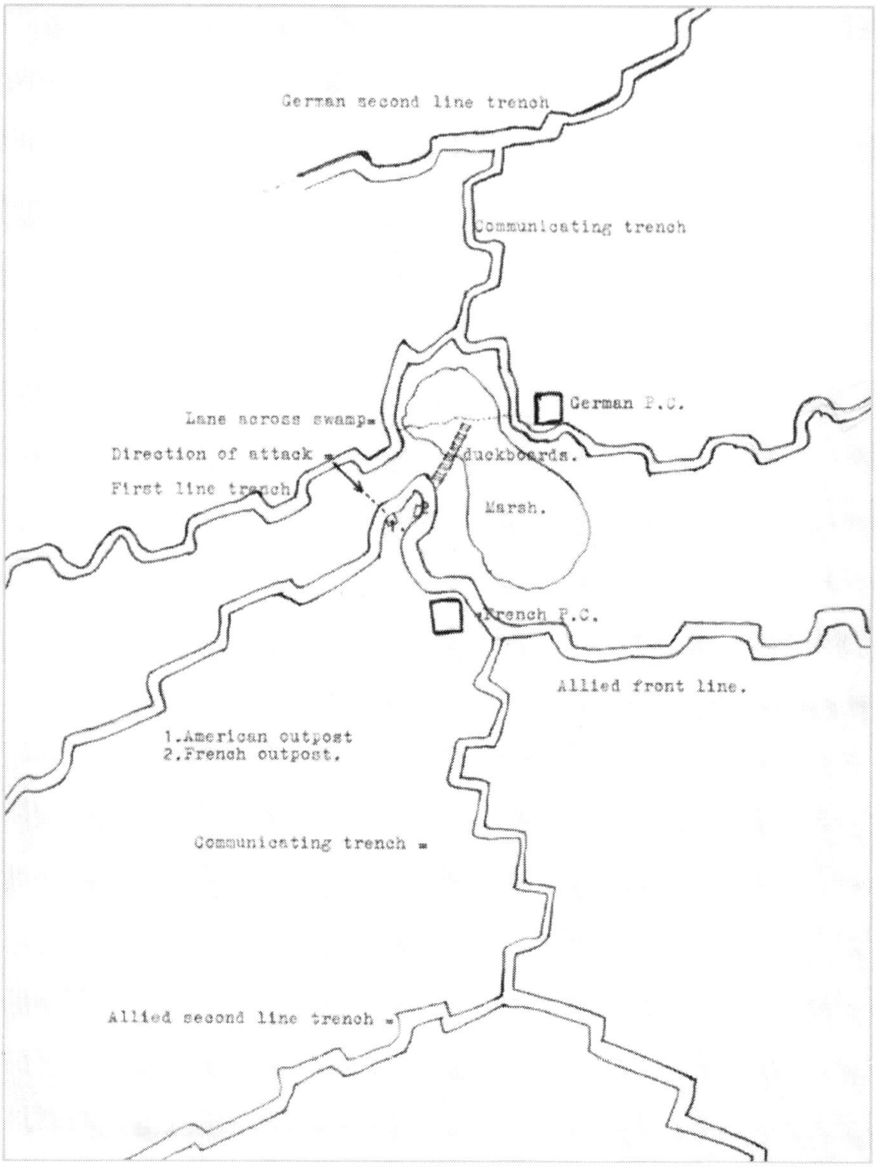

Rough military sketch of the terrain

The following little incident, the details of which are as easy to follow as they are to relate, actually happened; it was a big thing for the participants thereof but, inasmuch as it did not alter the war map and had no importance – judged from the general scheme of things - it naturally never appeared in any of the "communiqués"! Now that the ubiquitous Censor has been relegated to the dim past a rough military sketch of the terrain is permissible and will, I hope, enable the lay reader to follow the course of events without putting any strain on his brain cells. No attempt has been made to draw it to scale but it's perusal ought to facilitate an understanding of the melodrama as it actually unfolded. The participants in this big-little or little-big "show" must go unnamed; unsung and unknown heroes who "did their bit" in the greatest of world dramas. New-Englanders are rather notorious for their taciturnity; it's safe to surmise that the experiences of our 26th Division - the secret's out - will not be broadcast unless some outsider like myself - who took care of a goodly number of their wounded - opens the floodgates of his memory and makes them known!

On a certain foggy, damp, and extremely disagreable night in February 1918, between the city of Soissons and village of Vailly, a small outpost, guarded by a handfull of our youngsters, was trying to keep warm while it protected the bit of no-man's-land in front of it against a possible surprise attack. Leaving a sentinel in advance, the other soldiers were either huddled together in the upper room of the dug-out or peacefully sleeping in the lower chamber. No attack was contemplated on our part, and "military intelligence" had not observed any activity on the part of the Germans suggestive of an impending offensive. Ordinary sentinel duty was sufficient for the time being as this sector had become "stabilized", since the successful battles of October 1917 had enabled the gallant "Poilus" to occupy the entire length and breadth of the famous Chemin des Dames.

The front line, here as elsewhere, undulated back and forth, no-man's-land varying in width from a few yards to a mile or more. A marsh or swamp is always carefully avoided; no trenches can be built through them, and troops would be criminally negligent of the most elementary military tactics were they to make their stand with impassable ground at their backs. Field Marshall von-Hindenburg's fame was established at Tannenberg when he so brilliantly backed the Russian army, commanded by Rennenkampf, into the Masurian swamps where they had to surrender. Whenever the battle line became stabilized near swampy ground, both belligerants either retired behind or pushed well forward beyond the weak point, in order not to hamper freedom of movement or weaken lines of retreat. In a certain portion of the Aisne valley for instance, there exists a marsh, twenty miles long and from three to five miles wide. The German lines were stabilized on the north, and the French on the south of it. During the open season, officers and men of both armies very frequently went duck hunting. If the mallards were plentiful, the enemies avoided trespassing on each other's preserves and returned with wild fowl for all; if the duckies refused to be sacrificed to voracious military appetites, the divers' hunting expeditions "potted" each other instead!

To the right of the American outpost stretched a small, oblong, marsh, impassable during the rainy season; the German line passed well beyond it, leaving an open space two miles long and less than a mile wide. Running diagonally across the marsh, near it's north end, and at a point approximately two hundred yards from the Allied trenches, was a narrow and rather precarious path which was used at night-time by Fritz as a short cut to his battalion post-of-command. When the night was very dark and the enemy walked carefully, the path was reasonably safe; more often than not, they were spotted by a vigilant sentinel and paid the price of their temerity!

The particular night we are referring to has already been stated to have been foggy, damp, and extremely disagreable. Do not infer from this repeated statement that Sam Amex and his companions became drowsy and finally succombed to sleep at their posts; if they had, this story would never have been written or even thought of! Sam's eyes were quite open, and he faithfully and continuously tried to see beyond his own barbed-wire entanglements - an utter impossibility! He strained his hearing capacities to their utmost and heard all sorts of strange sounds which he endeavored to analyse correctly; he grew apprehensive a hundred times - and heaved as many sighs of relief - nature is never quite silent, even on the battlefield! At regular intervals, the sargeant in the dug-out sent a patrol outside, in order to make certain that all was well, and none of the soldiers in the upper room were allowed to even so much as doze off for an instant. In other words, our boys were doing their full duty. The adventure which befell them was an every day occurence all along the line. Their comrades had played somewhat similar tricks on Fritz more than once! Unwounded prisoners are being constantly brought in. They are useful in a variety of ways and, when important information is desired, "third-degree" methods were not unknown at German H.Q.!

That the local enemy "Kommandatur" desired some sort of information regarding the units and strength of our troops in this particular sector, is evidenced by the fact that a special patrol was sent out to bring in the U.S. outpost on the aforesaid foggy, damp, and extremely disagreable night in February 1918. The wet ground favored a noiseless approach underneath and between the wire entanglements, and the fog rendered them invisible. Alternately creeping, crawling, and pausing, like stealthy Indians, four Germans were able to pounce upon and secure our Sammy before he could utter a cry or discharge his rifle. As many a brave soldier had, in other raids and in the face of certain death, shouted a warning to his comrades, our struggling but now helpless Amex was, none too gently gagged by his captors. Having thus disposed of number one, the German patrol retired a few yards with their prisoner, leaving one of their men in the position of the ex-sentinel, and wearing the latter's helmet and trench coat. Amex number two, leaving the dug-out a few minutes later for his "check-up" round, met a similar fate! Patrolman number one having failed to return on time, a second soldier was sent out in the trench to investigate; he likewise was promptly pinionedand gagged. In his anxiety to discover what was wrong in his small section of trenches, our sargeant forgot to awaken the balance of his outfit before starting out on a personal tour of investigation. He was met at the entrance of his dug-out and, because the Germans were in a hurry to finish the raid and report to Battailon H.Q., they put Sargeant M..... temporarily hors-de-combat by the simple and most effective device of hitting him a sharp blow on the solar-plexus with the butt-end of a rifle! Leaving the four captives, solidly gagged and pinioned, under guard, the rest of the enemy patrol swiftly but silently rushed the lower room of the dug-out and soon had it's bewildered and still sleepy occupants powerless to either defend themselves or give the alarm to the next outpost, manned by French veterans.

Luckily for the ending of this story, Alphonse-le-Poilu, foxy old habitué of trench warfare, had heard the very slight disturbance incident to the rushing of the dug-out. Past experience made him sense a surprise attack and the fact that not a single shot had been fired, no hand-grenades had exploded, and no outcry had been heard, told him in no uncertain terms that all was not well with his new war-comrades. With feline agility, he noiselessly approached the Amex dug-out and listened intently for a few seconds! A guttural, rapidly smothered laugh and the word "Kommandatur" reached his ears and told him the whole story as completely as though he had been an eye witness throughout the whole episode. The prisoner's guard failed to see him as he crept along the trench back to his own post, where he recounted what had happened to his sargeant. The latter, realising that there was no time to be lost, ordered Alphonse to report his findings to the post-commandant, where a hurried council-of-war was immediately held. To return and open fire on the enemy patrol would certainly entail heavy casualties among the Americans: it was decided best to meet cunning with more cunning and spring a trap on Fritz which, if successful would - without killing anybody - still kill two birds with one stone!

The reader may remember that running diagonally across the marsh, and quite close to the Allied trenches, was a short cut leading directly to the enemy's "Kommandatur". It was quite reasonable to assume that Fritz, confident that his surprise raid had been a complete success, would utilize this path in order to bring in the Amex outpost they had just captured to their Commanding Officer, as soon as the necessary preliminary formalities had been attended to. These, O unsophisticated reader, consist in searching captives for hidden weapons, removing and collecting all tags and papers which would reveal the unit the prisoners belonged to, relieving them of henceforth useless articles such as money, watches, knives, and tobacco, and then, most important of all, cutting off suspender buttons and removing pants-belts in order to discourage any hundred yard dash or mile run, which an ex-athlete might wish to indulge in! These unpleasant incidents take time, and the French had a full hour before them in which to plan their counter-surprise.

As the ground, though soggy, was fairly hard, and the marsh beginning to dry up, it was believed that, with the help of duck-boards, a right-angle crossing could be made from the French outpost to the German short-cut. The necessary number of volunteers was soon obtained, and the proper number of duck-boards rapidly but silently taken from the trenches and laid, in a straight line, almost as far as the path along which they hoped to outguess the enemy. The impromptu pathway completed the Poilus spread out on either side of Fritz's lane and, carefully concealed from view among the marsh rushes, grimly awaited the latter's arrival.

Once they have been searched and deprived of suspender buttons and belts, prisoners of war are never manacled unless they have shown themselves as eager to escape; in the latter case, unless they are needed for personal interrogation regarding strength and disposition of their own units, a thrust from a bayonet soon brings them to their proper senses! In the zone of the interior for instance, it was no uncommon sight to observe a hundred or more enemy prisoners walking along the highroads of France, escorted and guarded by three or four old "territorials" i.e. reserves or "pépères" as the French called them, who marched along with them, guns slung behind their backs, hands in pockets, and pipes in mouth! The German patrol was no exception to this general rule; they strolled along their narrow path with the small Amex squad in the middle of the column. The Yanks all had their hands in their breeches pockets, in order to keep the nether-garments from leaving their proper station! Feeling perfectly safe from any retaliatory attacks, no precautions other than observing silence were taken.

The entire patrol was allowed to pass between the double file of vigilent but unseen Poilus but, no sooner had the enemy rear-guard entered the human trap than the latter was neatly sprung! From either side of the path, a French soldier jumped as if out of the fog; each Fritz was disarmed and rendered powerless and voiceless by four husky arms and hands, and our astonished Amex's found themselves among friends - free!..... Yes, a few obstinate German heads were cracked with the butt-end - and perhaps a bayonet or two was given a warning lunge but, again, no shots were fired, no hand-grenades exploded!

Fritz, once he has been made a prisoner of war, rarely attempts to escape. He becomes a remarkably docile captive, eager to work, and obedient to every order. Junker discipline, rather than an inherent martial spirit, welded the young men into that marvelous war-machine that was the German army!

Back along the new duck-board path tramped the Allies with their crestfallen captives, the latter wondering how it had all happened! The duck-boards were carefully picked up, section by section, and brought back to their old positions in the trenches. It was hoped that the same trick - minus the preliminary capture of the Amex's of course - could be repeated again and again, consequently all traces of the late scuffle were carefully removed, including the two or three wounded Germans who had shown an early tendency towards insubordination.

On arriving at the French C.O.'s quarters, the prisoners were lined up for preliminary routine formalities while excited Poilus recounted – with many a roar of mirthful laughter – all the incidents of the evening. Said the C.O. with a merry twinkle in his eyes: "Messieurs les Américains, do me the pleasure and give the enemy the honor of searching them yourselves; you may possibly be able to retrieve some of your own belongings!" No second invitation was required; in a few minutes, tobacco, watches, trinkets, and money found their way back to right=ful owners! No more thorough souvenir hunters than our boys ever trod the soil of France! Having found and distributed their own respective belongings, each member of the erstwhile captive American squad removed any and every appealing article from the Germans. Iron crosses, blouse buttons, corps insignia, cigarette lighters, and watches exchanged ownership; many of these now doubtless form part of some miniature "war collection" at home!

Cigarettes Américaines! What Poilu had not smoked scores of them by this time? These were first handed around and smoked, some even offered to the erstwhile captors but present captured - a gesture very typical of the Anglo-Saxon desire to be magnanimous towards a beaten foe. Judging from the distrustful attitude and sullen demeanor of wounded soldiers belonging to the Teutonic race who were brought to Allied hospitals, the inherent belief in fair-play - second nature with the average American or Englishman - is rather unknown in Deutschland. A certain well educated, though boorish "unter-offizier", about to undergo an operation for a badly fractured arm, once asked the writer if he was going to be allowed to live through the anesthetic - apparently laboring under the delusion that he might be deliberately disposed of – via Ether!

Picture the glee with which our soldiers watched the belt-less and suspender-buttonless Germans being escorted to the rear through the communicating trench! It required the stern warning of their sargeant not to let out a few good American war-whoops or college yells as Fritz vanished in the dark to be gathered up in some prisoner's cage! Gloriously happy, though still somewhat dazed by the rapidly changing events of the night, our boys, after hearty handshakes all around and many: "dors bien mon petit, sleep well, my boy, thanks, old pal" interpolations, regained their proper station where they were allowed the luxury of a few hours off duty, in order to calm down and prepare themselves for the next ordeal. A fresh detail had replaced them at their former outpost, and they were free to sleep or "talk it over"; of course, they did the latter.

"Gee! but that was a close shave!" remarked Tim; "how the h..l did them guys get us so easily?"

"Pshaw! that ain't no such a much!" growled Johnny, "let's turn the trick on Fritz, first chance we get."

"Say! ain't yer glad them Frenchies was next to us?" said Walter, digging Tom Barnes in the spare ribs with his elbow. Barnes, who had on previous occasions expressed profound dissatisfaction with the system of Allied troop intercalations, refused to answer, inwardly thanking his stars that darkness prevented his buddies from seeing the flush of shame which he could feel mounting in his cheeks as he recalled his former criticisms.

"What's hurting you, Sam?" inquired the sargeant; "let's hear from you".

The sentinel, still smarting under the belief that he should have seen, heard, or sensed the approaching enemy and not been caught unawares, turned a sad but determined face towards the last speaker, and muttered through clenched teeth: "Wait 'till I get at them; just you wait! There'll be something doing then, or I'll know the reason why!"

"That's the stuff, Sammy!" shouted the entire squad; "Soak the Germans; we'll all be with you at the finish!".

The Great Melting Pot

Early Months of 1918 – March 23, 1918

They were all Americans, all doing their "bit" for the common cause of liberty; what mattered a few generations more or less spent on American soil?

The great melting pot!
by
George Skitarnow

The Ambulance des Alliés, as it's title implied, was equipped and operated for the benefit of the Allies – at least that was the interpretation put upon it's name by the chief surgeon - an American. The term "Ambulance" as used by the French army was meant to describe a fairly mobile, but small hospital which could be shifted from one sector to another as the vicissitudes of the gigantic battle-line demanded. Several of these, including the one in question, had originally been equipped and maintained by generous and wealthy women of various nationalities. The impression one gathered after a close study of the problem was that, while the government officially expressed it's gratitude for the generosity of the donors and the French "Service de Santé", or Medical corps, utilized these semi-amateur, pseudo-military hospitals, they were, in the main, somewhat of a nuisance in that, never completely under military discipline, invariably backed by some more or less powerful politician in Paris, they did not harmonize with the stern necessities of warfare. The question sometimes arose as to whether such an "Ambulance bénévole" was maintained primarily for the self-gratification of it's socially ambitious donor, or for the alleviation of suffering wounded soldiers! Ostensibly, the "Médecin-Chef", or Commanding Officer of these "Ambulances", was in supreme authority; actually, he bowed to the whims and caprices of "Madame la Directrice" who, often enough, could summon a general or two who gallantly did her bidding and relegated the poor French surgeon to the role of a rubber stamp! Thanks to the vicissitudes of the battle-field, a few of these "society hospitals" rendered real service. Much diplomacy was needed to avoid friction and misunderstandings. English, Irish, Scotch, and American nurses, some of them professionals, others voluntary amateurs altruistically serving the cause and often eventually outdistancing the trained nurses in skilled attention to the wounded, frequently indulged in quarrels which had to be straightened out; French surgeons, with three years' of war-wounds experience, not infrequently questioned the ability of their American colleagues and were not averse to attributing any surgical failure to the "systême Américain". The chief surgeon vividly recalls an exciteable ex-football player, a towering young Doctor from Pennsylvania, warning him that "the next time any blankety-blank French surgeon came into his ward and changed any of his dressings, he would use him as a football!" Translated for the benefit of the French Army surgeons, this dire threat immediately put an end to over-zealous interference on their part! After all, one could hardly blame their apprehension. After three and a half years of constant warfare, poor France had almost been "bled white" and conservation of man power was a dire necessity with her people. How many of us here in America realise that, had our own losses been proportionate, our casualty list in killed, wounded, and missing, would have reached the staggering total of ten million men! Through painful and costly experience since August 1914, French surgeons had evolved a standardized method of treating war-wounds which they clung to with all the tenacity of their race. Add to this the fetish which they make of authority and tradition, and we have ample explanation for the many disagreements which cropped up, not only in the Medical Corps but among Combat Units as well.

When, in the early months of 1918, our 26th or New-England Division quietly slipped in and occupied a portion of the Couvrel sector, north of the Aisne river and bordering on the famous Chemin des Dames, they formed part of the French 11th Army Corps and literally rubbed elbows, company by company, with their new comrades-in-arms. Proper steps were at once taken by our Commanding Officer to receive our "Amex's" - favorite nickname overseas for the American Expeditionary Force - and give them a touch of home life. Our hospital boat, or "péniche" as the natives called it, fitted out with seventy bunks for severely wounded and capable of carrying an additional hundred and ten slightly wounded or "walking cases", had been in the habit of utilizing the Oise river canal down to the city of Compiègne, and then proceding up the Aisne river as far as Soissons, where our 26th Division

Headquarters had been established. The question of transportation thus became an exceedingly simple one - until a real battle developed!

As this particular sector was only to be temporarily occupied by our troops, it had been judged more expedient not to have any of our own Medical Corps' formations in the zone of the lines of comunication. A wounded American received first-aid in his regimental aid-station and was then sent back to a French Evacuation or H.O.E.(hopital origine étapes) for further treatment. As the Ambulance des Alliés was operated under the administrative supervision of the French Medical Corps, the transfer of wounded Amex's to our formation presented no formidable difficulties.

When the first group of wounded American men, including several junior officers, arrived in ambulances at the boat landing, saw the nurses in their smart uniforms and heard the English language spoken, after spending several days in a French hospital, they burst into tears and sobbed as though their hearts would break! Not that they had any serious complaints to make regarding the treatment they had received at the hands of French surgeons or orderlies, but it was so good and comforting to hear the mother-tongue spoken again and to have one's pillow and sheets smoothed out by a sympathetic nurse whom one could converse with without the use of the sign language!

A perusal of their "diagnosis tags", followed by short chats with one after the other of the first fifty or so American admissions, suggested the title of this story. Italy, Sweden, Russia, Poland, Armenia, Ireland, and the French province of Canada were all represented - often in the same company! A sturdy descendent of the Pilgrim fathers was flanked on either side by an offspring of the Ghetto and a Swedish emigrant; a Polish factory boy tried to converse in broken English with his neighbors, a French Canadian on one side, an Italian on the other; an irrepressible Irish lad flung Bowery jokes at an Armenian, the latter smiling his polite misunderstanding! They were all Americans, all doing their "bit" for the common cause of liberty; what mattered a few generations more or less spent on American soil? "We tutti have to crossa bigga da oceano soma time!" said Tony Damiano - and he was right, of course!

Ah! the proud look on the young men's faces as they, one and all, vouchsafed the information that they were not drafted but had volunteered! That the Middle-West is the great melting pot in America for the downtrodden masses of the old world, is a fact long admitted by all, but that this particular division, recruited in the New-England states, should contain the same alien elements in it's cadres, was most interesting and significant.

Beppo Mangiafava, called Maccaroni by the nurses, was as handsome a specimen of southern Italy as one could wish to see. Curly black hair, swarthy complexion, and large, luminous brown eyes reminded the historically inclined that Hannibal's legions had, not once, but many times, overrun southern Italy as well as Sicily, leaving the unmistakeable imprint of their passage on the features and complexions of the original natives. His smile was almost a perpetual one, showing a double row of marvelous ivory-white teeth. Beppino was, in truth, an artist's dream of a perfect model. His birthplace was "San-Giovanni di Cosenza, provincia di Calabria", his country: "me Americano, voluntario! Me landa en America at ten-a-year - quando avvevo dieci anni, si Signor; me liva en Lawrence - da-Mass, do evvyting - shina da shoe, worka da shov', sella da banan! Guerra, war, he come, me say: "Beppo, you musta fighta for ma new country"! Ecco, me soldier boy, var' fina ting: me wounda in da leg by big Germ': me killa da son-of-a-gun, no care go-dam, killa some more soma day!" Good old Maccaroni! In spite of his queer handling of the English language, his comrades respected and treated him as one of their own – as a pal.

Maddy Archanbault, lineal descendent of a long line of "voyageurs", was following the career of a lumber-jack in the Maine woods when the call came to him, and he enlisted in a nearby village. With his knowledge of "habitant" French, which managed to be understood wherever he went, Maddy had, immediately upon landing in France, become the unofficial interpreter, go-between, and purchaser of "extras" for his company mess. A quiet, modest, retiring sort of a lad was our Archanbault yet, on his breast was the coveted Croix-de-Guerre which a French general had pinned on him for conspicuous bravery! His face and arms had been badly peppered by shrapnel - which made the boy sometimes wonder whether his best girl at home, up in the Maine woods, would still favor him with her smiles when he returned. Good old Maddy! If Angéline is the right sort of girl, she will fasten her adoring and admiring eyes on your military cross and feign not to see the absence of part of your nose and upper lip! Bon courage and good luck to you! With Duplessis, Lambelle, and Girault - all good old French Canadian names - you formed a quartet of sterling fighters: would that more of your kind could slip across the border and enlist in our army!

Iky Rhozinski, typical product of any large city Ghetto, felt the importance of his new role in society! Gone were the sloping shoulders, pussy-foot gait, and shifting eye - inherited traits which are the natural result of centuries of persecution. Iky stood like a flag-pole and looked you straight in the eye; he had been wounded "in line of duty", suffered uncomplainingly - a rare thing with Israelites in general - and was secretly very proud of his bandaged and splinted limb. He felt that he was indeed an American - and his comrades treated him as such. Even the unexpected visit of a French military Rabbi who was able to converse with Iky in Yiddish, failed to elicit any sneering remarks or critical comment. Rhozinski was a "regular feller".

To Walter Brodowski, a Catholic, Iky should have been anathema, for there has never been any love lost between Catholics and Jews in unhappy Poland. The two frequently conversed together, now in broken English, now in Polish, apparently on the very best of terms. Forgotten - at least for the moment - was the veiled contempt of the one, and hidden fear of the other; comrades-in-arms, all! Walter - he confided to me that he had been baptized Stanislas, but had found Walter more comfortable for school purposes - was a typical Slav - tawny hair, high cheek-bones, low forehead, dilated nostrils, large mouth, and sad blue eyes. Several years of premature factory life had given him that dull, hopeless look so familiar to settlement workers. He was doing his manly best to overcome the old handicap, but it still lingered in gait and facial expression. Six months from now, if all goes well with him in the meantime, Stanislas will be completely emancipated from the old fetters, which had been prematurely crushing and ageing him at home. He will be a real man! Just what his attitude will be towards his pre-war "job" when he returns, is one of the problems - probably the most momentous one - for our sociologists to work out! Where are we going to get our future factory slaves from? Is a man who has breathed God's fresh air, slept under the starry heavens, and faced death unflinchingly, going to return to the States in order to again punch the time clock and be known as plain number 974, d....d Pollak?

Of course, Timothy O'Neil had freckles, carroty hair, a stub nose, and "pug" chin! Tim, enlisted belligerent in B.Co,Nth Infantry, entered the hospital with mitts clenched and an angry gleam in his eye! Shot through the left foot, with the wound badly infected and needing immediate surgical attention, Irish Tim announced to all who cared to listen to him that it would take ten nurses and six surgeons to get him into the operating room! "Tim, my boy," said the surgeon quietly, patting him on the shoulder while keeping an eye on clenched fists, "you will have to have your foot drained if you don't wish to loose it; take these crutches, walk with them to the operating room, and climb on the surgical table yourself." Blue eyes looked into equally blue eyes for a few seconds; then a sheepish grin slowly spread over the micky lad's face as he picked up the crutches and said: "Yes Sir!" From that moment, Tim became the star patient of his ward! Full of grit and determination, fairly bubling over with native wit, he constituted the only necessary tonic, disbursing this commodity in continuous heroic doses! My last recollection of Tim, before a hurried evacuation to a safer zone [that] brought our friendship to an untimely end, was when, leaning on his friendly crutches, he played ball with Iky - also on crutches - in the hospital yard before an admiring crowd of convalescent English, French, and American soldiers. Both were good pitchers and catchers; both stopped overhead balls with a crutch, and nobody sighed for a pennant game that day! How quickly a scene can change during war-time! That same courtyard, a few days later, as a result of the German onslaught of March 23rd, 1918, resounded with the daily and nightly rumble of ambulances bringing in desperately wounded men to us from the nearby improvised lines. From a comparatively quiet "society hospital", the American surgeons were suddenly called upon to operate, day and night, upon an apparently unending stream of head, chest, and abdominal wounds, wounds of such immediate danger to life that the poor victims could not have been carried to the rear of the battle-line without having first received life-saving surgery. Such wounded soldiers were classed as "intransportable" and operated upon in the nearest hospital available, manned by competent surgeons. We, the handful of surgeons and French orderlies who stuck to our post, provided the necessary hospital facilities during this tragic emergency. Ah! the sublime joy which comes to a surgeon as he saves lives, lives, and more lives! Souvenirs! souvenirs! how they crowd into the picture-; let us return to our melting pot!

Yan Jansen – no need to tell you that he was of sturdy Norse descent - had grown to be a famous ski jumper 'way back in Minnesota; he was also an expert lumberman and had been coaxed to the Maine woods by the lure of war wages and a foreman's job. Neither job nor wages had been able to still the great call to duty and adventure; Jansen had become an excellent ambulance driver, assigned to a French division which occupied a nearby sector. With scores of other fellow-Americans, he had been drawn into the desperate battle, finally receiving a severe shell-fragment wound as he was cranking his car for the twentieth trip from first-aid station to our hospital. Of his Scandinavian origin, only the name and physique remained, for Yan had finished high-school at home and now spoke French with a fluency, which made him the envy of his less linguistically inclined comrades.

John Arda, handsome, black-haired, and dark-eyed Armenian, had given up peddling imitation Oriental rugs in Boston in order to show loving gratitude towards his new fatherland. This was truly a holy war for John: his people in Armenia had not been heard from in two years, and he had given them up for dead - or worse! His face would light up with long suppressed passion as he recounted the tales of unprinteable atrocities committed upon his people under - in his opinion - the guise of "Kultur". Just as positive was he that he knew at whose door in Potsdam lay the real blame, whose were the brains which had engineered the wholesale Armenian deportations! Though wounded, Arda felt that he had only just begun to even up matters; he was eager to meet Fritz again and again on the battlefield!

As for the American-born Sammies - genuine Yankees most of them - they were in France because the honor of their fatherland had been flouted and trampled underfoot, and they had enlisted in order to right a hideous wrong. Splendid, lovable fellows, all of them, many were mere children in worldly wisdom: a few of the older ones more or less sophisticated. With their strong though boyish-looking faces, their manifest intelligence, and splendid physique, they were the admiration of their French comrades. With the help of a smile, a handshake, and a cigarette, new and mutually valuable international friendships were soon formed, and it was a joy to see Sammies and Poilus on a sunny afternoon - rare treats, alas, in northern France - basking out in the sun and carrying out all sorts of impossible conversations in Frankingle or Inglefrank - depending on whether French nouns and verbs were being anglicized, or English nouns and verbs frenchified!

What a liberal education for one and all of them; what a rare opportunity for free exchange of thoughts! After this hideous war is over, will there evermore be such a thing as a d....d foreigner - atleast among the Allies? Will not thisgreat cataclysm which has befallen the civilized world - save the mark!, bring about a better understanding among the people who have been gasping and struggling in our great melting pot? If it does, we, at home or in the battle-field, will almost feel repaid for the suffering we are enduring and the sacrifices we are making in order to see the struggle through to a victorious end!

My Pap he cry lika Hell

June 1918

In very broken English the boy answered all of my questions.

My Pap he cry lika Hell!

My short tale of the great melting pot ought to give the average reader an inkling of the many races represented in the American Expeditionary Forces. One often wondered why there should be so many "foreigners"; where all the real Americans had disappeared to when we declared war against the central empires and ordered draft registration. Only after returning to civil life was one able to get hold of official figures - and they rather dampened one's belief in native patriotism! The Provost Marshall General's final report did not mince words; if he could publish them, why not pass them on for the reading public to ponder upon? Among the draft registrations reported for desertion, 3.1/5% were American born, and only 1.4/5% foreign born. Only 20% of the population of the United States is foreign born, yet one third or 33% of the overseas fighters were foreign born. Of the total killed in action or dying from disease, 33% were foreign born. One half of all the workers in war industries were foreign born, and these same "foreigners" always oversubscribed their Liberty Loan quotas! Of course, one can think up any number of attenuating facts which would tend to modify these startling figures. Many of these aliens were eager to help in winning freedom for the countries of their birth. Poland, Croatia, Checho-Slovakia regained their independance as the direct result of Allied victory. Foreign-born individuals are more readily coerced, herded, or threatened than are free-born Americans. Whatever the dominant motive might have been in the individual, the fact remains that the casualty list proved that one and all, no matter what country or race they came from, fought bravely for their new country. Many of these young aliens had parents in various parts of Europe, and one can only imagine how heartrending it must have been for them not to be able to make contacts, to hold family reunions during the war. Requests for leave of absence were invariably refused unless one had been in France for at least six months. Crossing any frontier, even into Allied Italy, was out of the question.

One early morning in June of 1918, my Adjutant reported that an infantryman had walked into the hospital kitchens and asked for food. Suspecting that he might be AWOL or, worse still, a deserter, he had been fed and then placed under arrest pending a preliminary inquiry regarding his military status. Hurriedly donning my uniform, I walked over to the HQ building and sent word to have the guard bring in the suspect. A young Italian lad, barely twenty years old, was marched in. Of all the unmilitary specimens I had encountered, he certainly was one of the worst! His uniform was filthy, rumpled into myriads of creases, spirals torn and sagging over his ankles; all the buttons of his tunic were missing as were also the collar insignia. Even his overseas cap was missing! In very broken English, the boy answered all of my questions. He belonged to an Infantry division, which was undergoing intensive training in the Zone of Supplies or Communications Zone, therefore he had not deserted under fire. He had slipped away at night-time and taken a week to walk from his billet to the outskirts of Paris. For every meal which French peasants had given him, he had cut off a button and presented it as a token of gratitude. His overseas cap had paid for a day's ride in a vegetable cart, which had deposited him near the Bois de Boulogne. The driver had pointed towards the woods and shown him the way to the Hospital Américain, so here he was. He had not, however, come directly to us but retraced his steps towards the city for, to him, a paramount duty, an all-important meeting. Let him tell you the story in his own words:

"When draft he come, I shina da shoe in Pittzaburgam, make good mon. Boss shoe man tella me me have to be soldato, beat me up I no join subito, quick. Whata me could to do? Boss say same ting to other Italian boys; Santa Madonna Mia, we alla join! My fadder he live in Milano; me say myself, mebbe me can see him if I good soldato, why not? I go to place Boss show me for to take draft. Big soldier ask me many questions, write all same on card, pass me to big room full of other boys. Other big soldiers yell "taka off your clothes". I taka off my coat, my vest and stand still. "Take off your clothes" yell biggest soldier; he grabba my shirt and shaka me laka nutting at all. I shame myself when the clothes she is all on floor and I stand like picture of Bambino Santo; I never see myself before like I am made an I fella funny all over me. The other boys is same, all no clothes, all looka funny, all feela funny. Soon come more big soldiers and one looka my eyes and aska me can I read big letters on stiff paper on wall, and canna me read little letters on paper, and he shutta one my eye, and anodder my eye and he write on card and give me da card and tella me go to nex big soldier. Thees man he push bright strumento ina ma nose on one side and ina ma nose on udder side with bigga shine behine me and funny roun looka-glass with hole in middle tied to hees head. Then he looka in ears, see if they dirty, maybe, then maka me stan corner of room and ask me can I hear what he say. Sometime I hear, sometime no hear, he speak so soft, mebbe he voice tired? Da dentista he count ma teeth, he stick ma teeth with piece of iron and say O.K. pass on. More soldiers looka me over; I get socked on chest, punched in belly, rubber hose with hard end she is put on me for to listen to my heart. Then dottore he putta da hand on one side downstairs and aska me to cough and putta da hand on udder side downstairs and aska me to cough. Nudder dottore he maka me turna me back and bend down; me shame myself, me face she is red like Chianti wine but, per Bacco me soldato now, me obey.

In camp me good soldato. Soon we told we go to France. Quick I write mio Papa, tella him I go see him in Milano, have fiesta by him. Da big boat take us over she not so bad; me cross in steerage before! In France we told must learn be better soldier before fight da Boche. Much training, much marching, much work in trenches, much maka-believe battle every day. Soon I ask top-kick can I go see ma fadder in Italy. He laugh, da son-of-a-gun and say sure, go see da Colonel, he giva you ticket. I go to H.Q., try see Colonel. Adjutant aska me how come. I tella heem wanna see me fadder in Milano. Adjutant he say nuttin, write piece paper, fold him up so no can see what he written inside and tella me take him da paper to ma top-kick. I peela potato all day and sleep in guard house all night for ona week; I no see ma fadder in Milano!

Plenty week go by then, ona day me get letter from me fadder in Milano. He say he got ticket for Paris and tella me when he arrive and where me can see him. I go to top-kick tella him what da letter she says. He says, ya damm guinea shut ya trap before I slap you in solitary. Me fadder he is arrive in Paris la setimana scorsa, da nex week. Me slip outa camp, me walk an walk an walk. People all good to me; some talk little Italiano. Pay-day he no come in France, yet, so me no money. Me give one button, two button, all button; me give cap, but me get to Paris, find me fadder. Ole man, when he see me, he cry lika hell, he so happy. We kiss onda cheek we slappa on de back and cry on de front. Pretty soon fadder he say he got go back to Italy, ticket say so.

I walka back to woods, find hospital, get food and lock-up. Me no care what I get; I see me good fadder cry lika hell, he so happy when he see me!

Military infractions during war-time are drastically punished. Not having deserted under fire, I knew that Beppo, such was his name, would not be shot. Regulations demanded that the youthful offender be turned over to the Provost Marshall's office for trial and punishment. The influx of battle casualties at the hospital made it impossible to trace the romantic Italian's fate. He may have gotten off with a light sentence; I hope he was given an opportunity to wipe out the consequences of his filial devotion, that he came out alive from the war inferno and received an honorable discharge. If he did, shouldn't we all feel that it was worth the risk he took just to have his old daddy "cry lika hell, he so happy!"

Go-tha h..l, said the Parisians

Zeppelin Air Raids

Mid-March 1918

Unless one took part in the World War, the siren warning of air-raids or gas-attacks is as hard to understand as it is to describe. Picture to yourselves, if you can, a city dimly lighted as a war precaution, suddenly plunged into pitch darkness.

France

Go-tha h..l, said the Parisians!
by
George Skitarnow

Surgeon T.......of the Ambulance des Alliés Hospital, having landed in France five months previously, applied for and obtained the precious one week's leave of absence to which he was entitled. Five hectic months, battered around from pillar to post, welcomed here, buffeted there, through petty Army jealousies and misunderstandings, which he vaguely sensed but could not interpret; ordered to report to a Base Hospital near Bordeaux when his mission had been to obtain front-line information regarding the treatment of wounded, for the benefit of thousands of ex-civilian Doctors who were to follow him in France; miraculously rescued from oblivion, because a chief-surgeon was badly needed in a French advanced hospital where knowledge of the native tongue was almost a necessity, he had been on detached duty with the French 1Xth army.

Thanks to the intelligent cooperation and understanding of the French Commanding General, permission to visit First-Aid stations and Evacuation hospitals in that particular zone had been granted him. His particular mission - finally understood and encouraged by the new Chief-Surgeon of the American Expeditionary Force and his Headquarters Staff - was finished and the last chapter of his brief, concise book on "Surgery of the Zone of the Advance" had been mailed to Washington to be - as had it's forerunners - mimeographed and forwarded to Medical Officers Training Camps pending it's publication in book form.

Somewhere about the middle of March 1918, increasing air activity on the part of the Germans had been noted and commented upon. The British Vth Army had quietly replaced the French from St Quentin to Noyon, and it had been whispered - in strict confidence, of course - that the lines were very thinly held and could not withstand a heavy assault. That the enemy was seeking to weaken lines of communication was evident even to the military tyro. Railroad junctions, Gares Régulatrices - regulating stations - and amunition dumps were being systematically bombed at night-time. Heavy artillery fire was directed towards bridges, canals, and possible points of concentration in the rear or "zone des étapes" as the French called the beginning of the combat area. Officially, the sector was reported as "quiet": a few informative skirmishes or minor raids brought in a few casualties, but the surgeons were, for the most part "marking time".

After witnessing four successive air-raids in as many consecutive nights, our surgeon felt particularly "fed-up" with this most disagreeable feature of war, and it was with a sense of very genuine relief and satisfaction that he flung duffle-bag and musette, containing all of his best and shiniest "war-paint", into a compartment "réservé aux fumeurs" and climbed in himself as the train - poor war-worn train - slowly gave a forward jerk, then a backward shove and started moving southward from Compiègne to Paris.

He had set his heart on revisiting the blue Mediterranean and thawing out, after months of semi-frozen torpor in northern France. Coal and wood had been very scarce in his sector, and he had almost forgotten what a glorious feeling of warmth a glowing fire or steam-heated room could give one in mid-winter! At home, in the U.S.A., the papers had stated that there was one heatless day a week; in France, there had been seven - a decided difference in favor of home!

Puffing away at his pipe, the fervent disciple of Aesculapius closed his eyes and conjured up heavenly vistas of almond and cherry blossoms, flowering gardens, and luscious tangerines; he mentally reviewed the fascinatingly irregular coast-line running from Marseilles to Mentone, with it's islands, peninsulas, and bays; the glorious coloring, the coquettishly entrancing villas, villages, and towns - all possessing to a superlative degree that "come hither" aspect which bewilders the new tourist and makes him wonder whether he ought to get off at Fréjus, St Raphaël, Cannes, Juan-les-Pins, or Nice! Quite as a matter, of course, his thoughts drifted to that gifted writer and cartoonist George du Maurier, to beloved Peter Ibbetson, and divine Mimsie, Duchess of Towers! "Combien j'ai douce souvenance du joli lieu de ma naissance!"

To visit sunny southern France for the first time is unquestionably a treat to most tourists; only Peter and Mimsie and a few select mortals, flesh of their flesh and bone of their bone, can sense the sublime joy which comes to them when they revisit old and beloved haunts but "revenons a nos moutons": this is the story of a Gotha raid over Paris!

As the train slowly came to a halt at the Gare du Nord of the French capital at ten in the evening, the sickening moaning siren was heard, heralding another air-raid! Unless one took part in the World War, the siren warning of air-raids or gas-attacks is as hard to understand as it is to describe. Picture to yourselves, if you can, a city dimly lighted as a war precaution, suddenly plunged into pitch darkness. Doors and windows of hotels, restaurants, theaters, and dwellings instantly becoming invisible; lights which could not be hidden by shutters, blankets, or dark window-shades turned off and here, there, everywhere, from Montparnasse to Clichy and from Neuilly to Vincennes, the sad, woebegone sound of the sirens! Scurrying to safety in cellars, frightened civilians, screaming, crying, whispering, or laughing - a human gamut of temperaments and emotions -: mothers scolding or reassuring children, men cursing or giving orders, people lost in the impenetrable darkness asking their way or taking dubious protection in a doorway or porte-cochère, others trying in vain to decipher names of streets, squares, or to read house numbers; here and there, adventurous or foolhardy civilians and soldiers marching in the middle of a street as though nothing abnormal was about to occur. Suddenly, the hideous "wrang" of an aerial torpedo as it crashes through roof of building or on cobblestone pavement, scattering it's fragments in every direction, killing, maiming, destroying - a tornado in miniature! Artillery shells, with the exception of the "Big Berthas" which also fell silently, gave fairly accurate warning as to line of direction and probable location of grounding; not so the aerial torpedo whose fiendishly sudden explosion illuminated a section of darkness for a few seconds, to be immediately followed by cries of the wounded. As late as March 1918, most of the raids over Paris had been carried out by Zeppelins or "Gothas" as the French preferred to call them. Only during the last few months of the war did the famous German Fokkers attempt long distance flights over the French capital. In order to prevent the air-ships from flying low over the city and thus obtaining accuracy of aim and identification of objective, an ingenious system of heavy wires, strung horizontally from one stationary baloon to another, encircled the city. As night approached, the gigantic circle of inflated bags would rise majestically and in unison, taking station until the following dawn. Unverified gossip maintained that several Zeppelins and not a few enemy aeroplanes had struck these wires and crashed in consequence. As a means of air-raid warning, the French army had established gigantic drums on the ground, placed at an angle facing towards the enemy lines. To each drum were hooked up telephone wires terminating in receivers which a soldier of the signal corps kept attached to his ears. The vibrations of the Gothas engines were thus transmitted to the drum and the signal-corps man could telephone headquarters that enemy air-ships were on their way. Hence the sirens, giving the citizens time to turn off their lights and march down to their cellars before bombs could be dropped.

"Les Gotha! Les Gotha! Abritez-vous!" Take shelter! shouted the railway employes. Passengers rapidly poured out of the carriages and made for the exits. The customary exhibition of military passes and civilian tickets was dispensed with, and each individual bethought himself of two things: means of further transportation and a place of shelter while the raid was on!

A glance at the square in front of the station revealed the fact that no taxis were in sight; the underground railroad had stopped running and Parisians were free to use it's stations, passages, and tunnels as "abris" or places of safety. To run the certain risk of loosing oneself in the pitch darkness of the Paris streets was out of the question, besides which a two mile walk carrying duffle-bag and musette in hand was not at all appealing to an officer bent on getting a well-earned week's rest! One was, of course, free to remain on the streets during a raid, but nobody cared to take chances when not "in line of duty", so down the three flights of stairs of the "Metro" or subway went our surgeon, following the rest of his fellow-travellers, railroad employes, and the odds-and-ends of humanity indigenous to a railway station. The atmosphere in the lower platforms, the more daring of the Parisians walking back to their homes via the underground. Anybody familiar with the marvelous but intricate criss-crossing of the Parisian underground railway system will understand why surgeon T.......decided against tackling this Chinese puzzle!

Midway between the street and underground track levels, where tickets and newspapers are sold, is a spacious station room with several galleries leading to different subway lines. It was fairly bomb-proof, and some fresh air descended the stairs from above. Here stood, or walked back and forth, back and forth, a motley crowd such as one could only see in Paris during war time. Every imaginable uniform and all of the Allies were represented; even a few soldiers of unhappy Russia could be detected here and there in the throng, striving to make themselves as inconspicuous as possible - the Russian bear was not popular in France after the revolution of 1917!

Of panic, there was not the faintest trace! Occasionally, somebody addressed a scathing remark at the buzzing pirates flying overhead, and many sighed out loud the fact that they were missing a late supper or - worse still - a tender, pressing engagement! Total strangers harangued one another, the consensus of opinion invariably being that the Germans were "sales cochons, pas du tout polis, de vrais Boches, quoi!" dirty pigs, not a bit polite, in fact dirty Boches! How like them to have disturbed us at such an hour! The rest of the conversation, carried on in a dozen different languages, was jolly, animated, care-free! Dainty Parisiennes, caught in the subway station by the sudden cessation of trafic, chatted gaily with Poilus, i.e. French soldiers, Tommies, or Sammies: a soldier occasionally burst forth into song, the chorus or refrain immediately taken up by his neighbors. By that, I mean that more or less musical sounds issued from the medly of international throats present: the words...well...their absence reminded one of attempts to sing America or the Star Spangled Banner at home!

Presently, a Belgian "permissionnaire", or soldier on furlough, sat down on the floor of the ticket room, settled his back comfortably against the wall and began unpacking his musette. Out came an Accordion! As a loyal and devoted subject of brave King Albert, his first rendition was the "Brabancone", or national air of Belgium; wild applause greeted his unquestionably musical talent. An admiring circle immediately formed around him and, no sooner had he started the opening notes of a popular two-step than couples began dancing in the middle of the improvised dance hall - just as though they had descended the stairs of the "Metro" on that particular evening, not for safety's sake but for the express purpose of exhibiting their therpsichorean powers! Alphonse-le-Poilu grabbed Lizette-la-Grizette; Tommy's arm encircled Marie's waist, and Giuseppe il Bersagliere selected dark-eyed Madelon, who reminded him of Petrosina at home in bella Napoli! John and Sam Amex, still feeling rather shy in amigo France - especially after discovering to their consternation that the high-school curriculum on foreign languages failed to arouse any response when used on natives - indulged in a little "horse-play" shuffle of their own, heartily cheered by all for their efforts. A lone U.S. tar, furloughed from one of our transports, did a wonderful "pas seul", attracting an appreciative crowd of Irish, Scotch, English, and Anzac soldiers around him. Have we already forgotten that the Australian and New-Zealand divisions of the British army were called Anzacs? They were, and a finer lot of athletes and daring fighters never trod the poppy fields or mud - especially the greenish-yellow, slippy, sticky mud - of Flanders!

Waltz followed Polka; Polka was succeeded by two-step, one step, or no step at all; what did it matter! Such a panorama of the evolution of dancing was never before witnessed anywhere! A waltz could be hopped, bostoned, hesitated, or madly twirled until the participants fell, exhausted, into and onto the soft ring of humanity which was ready to receive them - and laughingly recovered their breath and equilibrium! Of drunkenness, of gross indecency, there was no visible trace. Individual flirtations were stopped or allowed to proceed as far as the assembled daughters of Eve wished them to - no further!

German papers of the Vaterland had been printing glowing accounts of the terror, which had supposedly entered the hearts of Parisians as the result of Zeppelin air-raids; the scene in the subway was a sublime refutation of the lies, which were spread broadcast for the purpose of deceiving semi-enslaved Teuton "leute" - the naif, simple-minded, gullible Fritz's who had been taught to believe in all official news!

As each successive squadron of Gothas was either fought off by the anti-aircraft defenses of Paris or, having slipped in in spite of them and discharged it's torpedoes, had turned back towards it's lair beyond the enemy lines, a period of calm enabled those Parisians who lived near the railway station to walk home leisurely, singly or in groups, laughing, singing, or gaily chatting as though they had just come out of Opera house or Theatre, after listening to a very jolly performance! By midnight, the departure of the last Gothas had been signaled. The street lights were turned on, and the subway officials obligingly resumed underground traffic until the last Parisian and the last Ally had safely reached his or her destination.

Gay, light-hearted, heroic Paris slept peacefully, while printing presses beyond the Rhine, obedient to the official news bureau and it's belief in "shrecklichkeit" issued "kolossal" accounts of terror-stricken Parisian mob, shrieking through the streets, fighting like furies in order to climb into railway carriages and be taken away, anywhere, far from the now totally destroyed capital! Truly it was to laugh "es war zum todt lachen!" as Fritz would say, could he but have had an inkling of the truth - at that time!

After a good night's sleep, Surgeon T.....of the A......hospital was able to secure his military ticket for southern France. The Traffic was normal, which meant that the train was crowded as usual with jolly English, Canadian, Australian, Belgian, French, and American officers, all on leave of absence.

The Mediterranean still had that indescribable deep blue tint: the blossoms, the warmth, the glorious sun, the vivid coloring - were all there, as of old!

The Promenade des Anglais still followed the graceful curve of the Baie des Anges at Nice and ..... but this is the tale of the Gothas!

The Omnipresent Censor

Somewhere in France

Tuesday, February 26, 1918

Many a fact of importance to the enemy might have slipped through,
not because of malice aforethought, but solely because of carelessness.
In a general way, just what does the enemy desire to know
concerning his opponents? Impending battles, of course,

The Omnipresent Censor!

Just what is permissible and what is forbidden in letter writing during war-time can only be learnt by using common sense and submitting each and every individual to a self-questionnaire. Many a fact of importance to the enemy might have slipped through, not because of malice aforethought, but solely because of carelessness. In a general way, just what does the enemy desire to know concerning his opponents? Impending battles, of course, shifting of "shock-troops" for the sole purpose of "going over the top" and attaining a certain military objective, strengthening of the Reserve forces, new batteries of artillery, massing of tanks, increased activity in the rear such as new ammunition dumps, and other combat supplies. In addition to these facts, the enemy is particularly anxious to know what the morale of the opposing troops is. Are the men dissatisfied, discouraged, worn out? The shifting of Divisions had to be kept secret as long as possible. Not only did we know that certain of our Divisions could only be used as Reserve or broken up for replacements, but it did not take long to conclude that four Divisions - the First, Second, Twenty-sixth, and Forty-Second - would have to be called upon again and again to assault the German lines. It was natural therefore that the enemy should be particularly anxious to know the whereabouts and intentions of these magnificent units. Time and again, they had to be thrown in line to save a critical situation. When, on the 26th of May 1918, the Germans surprised and overwhelmed the thinly held Franco-British lines guarding the Chemin des Dames and began to cross the Marne river preparatory to a march upon Paris, our Second Division - which included the Marine Brigade - was actually in billets in a reserve sector near the town of Mondidier. Hastily placed in trucks, the Division was rushed across to the Chateau Thierry sector where the French lines were in jeopardy. The inhabitants had closed their houses, packed up whatever they could transport, and were trekking heartbrokenly towards the south. Suddenly towards them came army trucks, by tens, by fifties, by hundreds, filled with our Doughboys, singing, waving flags, and indulging in "horse-play" as only American boys can do, even in the army! The psychological effect upon the inhabitants was magical! Shouting to one another "Ce sont les Américains! les Américains! Retournons chez nous; tout va bien!" - These are the Americans, the Americans! Let us return to our homes; all is well! And, as a matter of historic knowledge, within a few hours, homes were unlocked, shops reopened, and village upon village along the Paris-Chateau-Thierry road had resumed it's normal life - or had I better say, it's normal war-time life? The Germans had already, at Cantigny, been taught to respect the fighting value of our First Division; Chateau-Thierry, Vaux village, and Belleau woods made them thoroughly acquainted with the tenacity and superb marksmanship of the Marines. It may be quite true, as General Ludendorff writes in his memoirs, that our Second Division did not achieve any tactical advantage, but the fact remains that they, step by step, tree by tree, ravine after ravine, consistently beat their opponents back, and that the march towards Paris was permanently stopped. The wards of my hospital were full of wounded Marines for several weeks after Chateau-Thierry; we had ample opportunity to learn, at first hand, how stubbornly brave they had been, how with bulldog tenacity, they had cleared Vaux village and Belleau woods of their tactically better trained enemy.

All Military hospitals in combat zones were enjoined not to write about the units to which wounded belonged. This precaution was quite easy to understand. If large numbers of soldiers belonging to a shock Division were wounded, it could only mean to the enemy that that particular Division had been retired from the front and was in billets in the rear, getting replacements and training the latter. Names of cities, towns, villages, rivers, valleys, and hills of tactical importance had to be omitted from all letters. The entire battle front had been arbitrarily divided into Postal Sectors, which were shifted about with Divisions or Army Corps. While Chief Surgeon of the Ambulance des Alliés, Secteur Postal 204 was my address. When the Germans crashed through in March 1918, almost overwhelming the British 4th Army, the French hastily filled the gap, and my Postal Sector became No 6.

Wishing to give my colleagues in the States a glimpse of what I was going through, I wrote the following short article which was published in the *Journal of the American Medical Association* on March 30th, 1918:

Somewhere in France, February 26th, 1918.

"The censor permits me to state that I am in France, in the zone of the armies; that I am chief surgeon in a hospital attached to the X... corps, Y... army, in the Z... sector! I shall try, in fear and trepidation, to convey to you the information that twelve of our boys of whom two are officers - are at this moment sharing the benefits of this establishment with their French comrades. (These wounded belonged to our 26th Division - New-England - and this was their first contact with actual warfare. The division was occupying a sector near the famous Chemin-des-Dames and the different units were sandwiched between similar French units for training purposes but, as no American army hospitalization had been provided for them, they were taken care of in hospitals under French army control. After receiving first-aid, they were placed in an improvised hospital barge called a Péniche by the French and taken down the Oise river to a point one mile from the hospital where ambulances awaited them. How these boys did cry with joy when they saw our English and American nurses smiling at them from the river banks!) Needless to say, no mention of these wounded nor of the Division they belonged to appeared in my article!

"Mr Censor may even allow me to state that the hospital grounds can still boast of artillery emplacements, trenches and wire entanglements although the veterans of the French army are now busily removing these reminders of the terrific conflicts of 1914-1917 - which fact, if you only stop to think of it, indicate to my mind that the general staff is not paying much attention to Hindenburg's recent boast that he would dine in Paris on April fool's day! (Alas and Alack! Three months to a day after this was written, the Germans crashed through the Chemin-des-Dames, began their march to the Marne and Chateau-Thierry and forced us to evacuate the hospital after sending all our wounded behind Paris. Trenches were hastily reopened, barbed-wire entanglements replaced and batteries of 75's hidden in the woods of the Gannelong hill which rose just behind the hospital grounds!)

"When I think back to the strenuous days of Fort Benjamin Harrison (Indiana) and recall the "angor animi" which overwhelmed me when, as C.O. of my company of medical officers I shouted "Right front in line" instead of "On right into line", thus scattering my men all over the railroad track and causing our esteemed instructor to yell: "Look at you, look at you; just like a flock of sheep!". I realise that the scene has shifted quite considerably and that other matters, equally vital, now occupy our thoughts!

"Everything is topsy-turvy; as I write in bed, (because there is absolutely no wood or coal available except for cooking purposes and the old Chateau is a huge ice-box in winter-time) a glorious full moon illuminates the country around me, but heavy black curtains over my window and a tightly closed door, prevent me from gazing at it; no lights being allowed to appear from the outside for fear of attracting the attention of an enemy night prowler of the air. "Damn the moon", remarked an officer to me as we gazed skyward before retiring "this means a visit from Fritz to night".

A clear day means "good visibility" and, while I am strolling along the high road, enjoying the beautiful scenery, Fritz, sitting in the basket of the "sausage" stationary balloon back of yonder hill, is probably telephoning a Boche battery words to this effect: "Idiotic Yankee on high road; three thousand yards; give him hell!". Fortunately you hear the whizz of the coming shell, jump into the trench which borders the road and lie flat on the muddy bottom until the beastly thing has exploded. Other shells may follow the first one, or Fritz may only have wished to notify you that promenading along the highway was not a healthy occupation and you retrace your steps until a turn in the road hides you from those invisible but all-seeing eyes.

Aeroplane bombs give you no warning: you hear the faint purring of the engine, several thousand feet above you and recognize it's nationality by ...... beg pardon, Mr Censor, but the plane itself is invisible! (Now that the War is over it is quite permissible to state that all German planes had a slight but unmistakeable miss is the purring of the propeller: instead of a steady zoom such as the British and French planes gave out, the sound was somewhat like wurrh, wurrh, wurrh, wurrh.) You take to the nearest shelter and trust that the bomb will not make a direct hit. A terrific "wrrang" loudly informs you that the bomb has touched ground and exploded. As aeroplanes travel rapidly, if the first bomb has fallen near you, you are perfectly safe and crawl out to take a look around; if the explosion is a distant one, you crouch and wait – perhaps for your turn to come! Of course, if a squadron - or escadrille - of planes is passing overhead the hits may be quite close together. (One horrible night in March 1918, fourteen bombs fell in the hospital grounds in as many minutes, shattering windows but, fortunately, making no direct hit. The Germans were trying to destroy the Paris-Noyon railroad which paralelled the hospital a few hundred yards away. As for the size of the excavation caused by a single bomb, I took a snapshot which showed ten officers or nurses standing in a single hole, Only heads and shoulders appeared above ground!)

Protection against shell fragments is easily obtained. Tiers of sand bags, filled with earth or gravel, or two parallel sets of chicken wire, one foot apart, strung around the hut or ward and filled in with earth, keep you perfectly safe. If the bomb should happen to hit the roof...that's another matter! Hospitals are now provided with bomb proof dugouts and, if the alarm is sounded soon enough, patients and personnel can be placed in absolute safety even against a direct hit.

It is a long way from "pup tent" which we were taught to put up or take down in three minutes and forty-five seconds, to the wooden huts in which officers of the medical corps now live when in the combat zone! It's roof and walls are camouflaged; double chicken wire and earth surround it; doors and windows are tightly closed and curtained at night, as long as any light is burning. In one corner is the cot, surrounded by a three foot wall of sand bags, giving one the illusion of a bath tub! You remove a few clothes - none when the moon is shining - take a running jump and clear the sand bags - excellent exercise just before retiring! Having accomplished this acrobatic feat, you place your gas mask and helmet where you can instantly grab them, test your flashlight to see if it has not gone "dead", pull the woolen "crusader's head gear" which loving fingers have knitted for you over your head, and settle down to dream of enameled bath tubs with hot running water, Castile soap and Turkish towels! Instead of the rumble of elevated trains or honk-honk of Automobiles on the boulevard, the dull boom of heavy cannon, or short angry bark of anti-aircraft guns, soothe you to sleep."

Cherchez la Femme

Friday, March 22, 1918

In sophisticated and logical-minded France, whenever a man gets into serious trouble, within or without the law, it is quite customary to shrug one's shoulders and murmur: "cherchez la femme" – look for the woman!

Cherchez la Femme!

In sophisticated and logical-minded France, whenever a man gets into serious trouble, within or without the law, it is quite customary to shrug one's shoulders and murmur: "cherchez la femme" – look for the woman! Sadly, one usually finds that a woman is at the bottom of the mess! It's so very, very hard for the darlings to keep a secret, to maintain a discreet silence, even during war-time when an unguarded sentence may make or mar a campaign or battle.

Having watched the blossoming romance which had sprung up between a wealthy expatriate American woman and a gallant French general, none of us were surprised when the engagement was officially announced and all officers present were invited to a grand pre-nuptial dinner. Three generals, as many colonels, and a baker's dozen lesser fry sat down to eat and drink to the happiness of bride and groom. The former gave detailed information regarding the wedding trip, which was to take them to Montebello, Italy where her spouse, so she said, had been appointed as commander-in-chief of the French troops in which a few months previously, had stopped the famous Caporetto disaster which Italians, quite naturally, do not to this day care to discuss. It was noticeable that the general himself, as well as his fellow officers, remained silent – but who can stop a woman when she has something on her mind, which will dazzle her audience and cause envy in the hearts of her less fortunate sisters! The monologue included a detailed description of the castello she was to occupy, the servants she would take with her, and the festivities she would inaugurate for the benefit of her husband's staff. And the general and his fellow officers still remained silent; the A.D.C. seemed ill at ease and frequently stole furtive glances around the table, apparently anxious to guage the effect of Madame's ebbulitions on the American contingent of officers. Dessert and the last toast disposed of, we all moved to the parlors for after-dinner coffee and liqueurs. Almost as though the climax had been a prearranged affair, an orderly entered and requested the A.D.C. to go to the telephone. Within a few minutes, he was back again, pale and agitated and hurriedly whispered a few words in the general's ear. The latter had nerves of steel. His face showed no emotion, it did not change color; hands and arms remained motionless. He stood still for an instant while we all seemed to be holding our breaths, waiting we knew not what for, but expecting some calamity, military or otherwise, then calmly walked to the fireplace, leaned his elbows on the mantle, faced his audience and, in the most matter-of-fact tone of voice, said: "My assignment to Italy has just been cancelled, and I have been relieved from duty with front-line troops!" Murmurs of consternation burst forth; Madame collapsed on a sofa and, still full of her social ambitions which she felt were to be thwarted, cried out: "But I shall be marrying you under false conditions!" The general did not reply to this sally; the hitherto gay gathering broke up and, as the A.D.C. was leaving the room, he beckoned me to follow him. As soon as we were out in the courtyard of the chateau and safe from any possible eavesdropping, he wrung his nervous hands, gesticulated wildly as only a thoroughly wrought-up Frenchman can do, and whispered: "It is incroyable, unbeleavable, Madame tried to send a long cable to American announcing her future husband's plans in Italy; the cable was stopped and turned over to G.H.Q.; what an indiscretion! We are ruined, done for, Limoge – and even our horses are at this moment on their way to Montebello – what a disaster, what an unforeseen calamity; what will the poor general do?"

The wedding took place the following morning in the little chapel of the chateau. Again, a brilliant gathering of military men with a few civilians who came up from Paris for the occasion. It all seemed like a sad anti-climax; nobody seemed happy, and all were wondering what was going to happen to the poor general's military career. He was at first "mis a disposition", i.e. ordered to await further orders from G.H.Q. Returning from Paris on the night of March 22nd, 1918, I entered the compartment occupied by the general and his wife who were, presumably, returning from their honeymoon trip. Beginning at the town of Pierrefitte, we ran into enemy air-raids. Tracks had been blown up at Creil, which was an important regulating station for military trains, and a number of French soldiers had just been killed by one of the bombs. Poor fellows, they were returning from a ten day furlough! The railway station at Compiegne was a sight! Not only had the German airmen succeeded in blowing up a huge gazolene depot, but a direct hit had demolished the station restaurant, killing the proprietor and a number of officers and soldiers. I was very kindly invited to share the general's car, and we began our slow journey towards the hospital, ten miles away. Military convoys everywhere; foot soldiers, artillery, munitions, quartermaster supplies of all kinds blocked the road, and it was early morning before we reached our destination. "Don't mention what you have seen to-night; the times are very serious, but we cannot discuss them at present." So spoke the general who, naturally, was better informed of what was really taking place than a mere Army Surgeon could be. As a matter of fact, this was the very night in which the Germans launched their gigantic battle, their battle for freedom and victory!

What must have been the general's thoughts that night, as he realised that he no longer commanded the troops of that vital sector, which was destined to see the final checkmate of the enemy onslaught! He made one or two appearances during the following week; his assignment was the care of some twenty-five thousand German prisoners, somewhere in France. So ended a brilliant military career – is it any wonder that the French persist in saying: "cherchez la femme!"

Censored Love. Love Letters.

March 1918, shared 14 years later

While vaguely wondering what I should do with them, I gathered the twenty alternating-colored letters. It seemed uncanny that they should have fallen, one by one, in precicely the same order that they had been kept in.

Censored Love.
Love letters
Translated from the French
and edited by
George Skitarnow.

Property of
Dr George de Tarnowsky
Colenel Med-Res,U.S.A.
30 N. Michigan Ave
Chicago,Ill

Fourteen years have elapsed since that memorable morning of March twenty-first, 1918, when the Germans began their desperate and almost successful attempt to separate the British from the French armies and throw the former into the British channel in utter rout and confusion! Dark days of anxiety were they, almost of despair, many of us wondering if, after all, we of the A.E.F. had not arrived too late to stem the tide! On detached duty with one of the French armies, I was at that time chief surgeon of a mixed hospital where American and French wounded arrived from the battle-line to the north of us, doing my bit while waiting for the bulk of our own troops to be drawn into the conflict.

Fourteen years have elapsed, during which period I have sacredly kept the mysterious letters [for] which I now feel at liberty to copy; they were too beautiful to remain indefinately in the privacy of my army locker, and, at present, their publication can harm nobody. They came into my possession one afternoon in March 1918 when, worn out by the strain of operations and the sight of moral and physical suffering with which we were constantly surrounded, I was strolling along a nearby picturesque and deserted forest road in northern France. Puffing at a pipeful of the soothing weed while vainly endeavoring to drive all painful thoughts from my mind, a sheet of blue letter-paper suddenly caught my eye as it lay in the middle of the road, unfolded, idly flapping in the wind. Similar sheets, some white, others blue, lay scattered in front of me, as far as the somber depths of the forest enabled one to see. Idly picking up the first one, a woman's handwriting revealed itself, the rapid scrawl of an intellectual; the second one, of white paper, was covered with a man's unmistakable chirography!

Some refugee, hurrying from the enemy, had undoubtedly dropped them out of a bag or bundle. Why did I pick them up? Why do we perform so many automatic actions in the course of our lives? As I gathered them together and placed them in what was evidently their blue, white, blue, white order, my thoughts naturally reverted to the atmosphere of tense excitement which still surrounded us.

Tears came to my eyes at the recollection of two girls, both under twenty, stopping in front of the operating room windows and leaning against a wall to rest. Taking turns, they had, for miles and miles along dusty roads, trundled a wheelbarrow on which was a single black trunk. The unique bit of luggage had fascinated me; it was so old, so dilapidated; it's four corners gaped as the result of the frantic packing it had received and only a stout rope prevented it from bursting open and scattering it's contents to the four winds! The younger girl began to cry, and I had turned my gaze up the road - in order not to cry - for strong men learnt to cry during the war; cursing was so inadequate to express the feelings which surged within one!

How much one can recall in a few seconds!

While vaguely wondering what I should do with them, I gathered the twenty alternating-colored letters. It seemed uncanny that they should have fallen, one by one, in precicely the same order that they had been kept in!

Blue, white, man, woman, boy, or girl - who on earth could have dropped them? Ah! What have we here? A piece of pink silk ribbon, partly untied, near the very last letter; a girl or a woman has lost her precious correspondence! Who is she; where is she; what business is it of mine, anyway?

The forest road I stood upon, skirted the base of Gannelong hill, the summit of which was the point of vantage we were wont to gain in order to watch the never-ending battle for the possession of Mount Renaud, a small eminence dominating the city of Noyon on our right. There was not a human being in sight; no birds, not even a squirrel to break the nerve-racking sound of cannon or the sharp, angry ta-rac-ta-tac-tac-tac of machine guns! Amid such gruesome surroundings, how could one expect to have a frantic damsel rush up to one imploringly: "Ah!, Monsieur! My letters, my letters! For the love of God and your sainted mother, return them to me; they are my honor, my...Ah! thank you, Monsieur, thank you!"

Nobody came; I stood as if rooted to the ground, holding the missives in my hands, undecided, intrigued, unwilling at first to read a private correspondence yet irresistably compelled to do so, hoping to find some name, some address which would enable me to return them to their rightful owner. To throw them down again was out of the question; who would pick them up again, and what use might this person not make of them? Revenge, blackmail, public scandal, anything might be the result. Any attempt to burn them up in the road would have brought a patrol - open fires were strictly forbidden - and precipitated a court-martial. A hurried glance at the first few confirmed the suspicion that they were love letters; distributing them in my various pockets, I retraced my steps and, in the privacy of my modest quarters, placed them in my locker where they remained to this day. No other eyes scanned their contents - or ever shall; this world of ours is infinitely small and...some handwritings are so easy to recognize!

One or two at a time, as privacy and opportunity offered, the letters were read, studied and carefully replaced in their proper order. There were no envelopes in the collection, therefore no tell-tale post-marks; no place or dates had been used by either correspondent although the man's letters frequently had

"Secteur Postal No 702" on them. This, in itself, gave no clue, as military postal sectors moved from one part of the line to another with the constant shifting of Divisions or Army Corps. Each letter began with M'Ami or M'Amie - dear friend, and ended in the same enigmatic way; identical intimacy and...discretion!

During the ten weeks in which our hospital was able to treat wounded, before the surprise Chemin-des-Dames offensive compelled a hasty retreat, many quiet, non-commital inquiries were made in surrounding villages or towns; no clue as to the possible identity of the writers came of them. They had completely vanished!

Somewhere in France, of this I was certain, a young woman was worrying her heart out over the loss of her most precious possessions, her letters and those of her lover. The bit of pink ribbon, to me, meant that the unknown man received one of the love letters, answered it and sent hers back with his - a wise precaution on his part, for dead men's letters sometimes have an uncanny habit of turning up where they are not wanted or cannot be understood, and life in front-line trenches is, at least, precarious! The only possible indiscretion was hers, in keeping them, therefore she was young, very young, and extremely romantic.

Having utterly failed in my attempts to find the rightful owners, the letters, pink ribbon and all, remained in my locker and were brought home with other precious souvenirs of the great period of idealism. Months after my discharge from the army, during one of my periodic moods of longing for some of the noble realities of life which the war brought forth, I opened my locker, read a few of my army orders, fussed with snap-shots of comrades, and fondled the impossible assortment of belts, buckles, spurs, and other military regalia which it contained. While thus occupied with grave yet fond reminiscences, my hand suddenly encountered a bit of pink ribbon, and I drew forth the package of blue and white letters - my unsolved enigma, as I used to call the bundle.

Why not make an attempt to complete the fragmentary story, to fill in the few gaps, in order to arrive at a better understanding of the lovers' state of mind, of their longings, and, especially, of their philosophy? Both belonged to the same school of thought, both corresponded quite frequently in the form of meditations or confessions, both indulged in self-analyses, tried hard to and apparently succeeded in convincing themselves that they were on the right path and, having done so, opened their souls in a series of precious and well-nigh inimitable reflections.

The lovers both suffered from the fact that their state of happiness was not a complete one; each in return realised that, by yielding to the irresistible surge of their harmonious souls, they had run counter to social customs, which bound them and amongst which they were, in spite of themselves, obliged to live! Notwithstanding the discretion which they maintained throughout their correspondence, one could readily sense the obstacles which prevented them from looking at the future without some misgivings. One of the lovers, the man, was not free! He had found happiness after much unhappiness; she, having given herself to her lover, was constantly asking herself where lay the truth in life!

Their meetings were difficult to manage; they were perforce, irregular, and very short, hence the extreme sadness of their partings - with death lurking over the battle-fields, each kiss might be the last one. Her first letter, though undated, unquestionably marks the prologue of the lovers' future intimacy. The handwriting is a rapid, though bold and precise scrawl, quite distinctive, easy of recognition, and impossible to disguise.

"M'ami! M'ami!" she writes, "how did it happen? I live over and over it, moment by moment and try to make the wonder complete, but the beginning will not come back to me! All I know is, that we were standing apart in the hall, and the next thing I remember is that we were in each others arms and your soft beautiful kisses were overcoming my face, then my throat, and then paused, for wonderful heart-throbs of joy, on my lips! My very life seemed drawn from me oh! angel of my heart, except for the ecstasy that thrilled through me till it hurt.......And then you said "good-night m'amie" and were gone. I held the door partly open, I don't know why, until the outer door closed. The concierge (Janitor who, in French appartment buildings keeps the front door locked from sunset to sunrise.) turned the lights out; the stairway was plunged in utter darkness! Then I closed my door too, and stood leaning against it with my cheek to it's cool surface, for a space. I could not think, there was no time-sense for me, I just lived in what had just been and would be forever an unforgetably wonderful experience. For hours later I sat with covered eyes, or stood here or there, as I walked about, leaning my head or my arms against anything I was near, for support, living again and again every minutest detail over - and always the ecstatic hurting throbs coursed through me! Night kept vigil with me and morning took his place, so both know my secret, that no one but you and they shall ever know. And I wondered if this miracle had affected you in at all the same measure and manner. I liked to think that you too were awake, dreaming about it. It seemed and seems like a dream, except that your kisses are still vivid and real - so I know that it was not a dream! Mon petit chou bien-aimé, my beloved little cabbage, I should not write in this way but I can't help it! You are stronger than I. I should like to pour out all the longings and yearnings that fill me to overflowing, but I must not. I think and live and think over all the precious moments! I want to know life - with you m'amie - I am ready, come and get me, soon, oh! very soon - hurry! hurry!"

Apparently following the "prologue" in chronological order is a long outpouring which fills three double sheets of her letter paper, written in the form of a meditation or communion with nature and self. It might well be entitled: "Elle s'est donnée", for the lovers had met again and....loved! The young woman is evidently in her parent's home in the country, and either near the sea-shore or some other body of water for she writes:

"Oh! melancholy! melancholy! You crush my very soul to-night; I know not whether I ought to hate or to love you!

The sun is setting and the water is on fire.... gold, purple, scarlet; all the warm colors intermingle, the lights are vivid and there is a beauty, a perfect harmony, as invisible strength in nature which grips and leaves me without spoken words! In the music-room a relative is singing l'Heure exquise - the exquisite hour!

"The sun has disappeared behind the hill; now comes the violet-colored hour, later on the fleeting moments of contrasts, of darkening shadows; purple silhouettes stand out from a sky which is still warm-colored...the waters appear to deepen in front of me while, above my head, amidst the tree-tops, can still be heard the sleepy chirping of birds!

"Indoors, the same voice is now singing Ninon! Ninon! what hast thou done with life?

"The hour of silence is at hand; an adorably beautiful star scintillates overhead, plays with it's reflection in the water, as though beautifying itself for the night!"

Can you not see her, even as I do in my mind's eye, sitting at the water's edge, knees drawn up under her chin and hands clasped together around her legs, questioning the infinite?

"Oh nature! nature!" she continues, "you are beautiful and I adore you, but do you realise how cruel you are? Is it not true that your beauty cries out Love! Love!, infinite, intense, tender and yet all-powerful love which creates harmony? Oh nature! you who are indeed so perfect, why don't you tell me what my senses, all of them, ask so fervently of you? Which road must I follow in order to be nearest to you, what human laws and customs must I accept or reject; what do you wish me to be in order to resemble you, to be harmonious with myself and yourself? Show me, oh nature!, show me the way and I shall follow you even as an obedient and confiding child and pupil follows his master!

"Only a night-bird answered my question; it's note sounded beseeching; the night is dark......I am crying!"

Poor anxious soul, how earnest, how sincere is her cry of distress! The first reading of this meditation caused a painful contraction of my throat, even as it does now - although I got to know it's every word by heart while trying to find it's author -, then I feverishly hunted for and found the man's answer to it. I wished to know what type he represented and, as the realization came to me that he, also, loved with the full strength of his soul, even as she did, my heart filled with sympathetic joy! It was not, it could not have been, one of those superficial, quickly forgotten, war-engendered flirtations or liaisons which burst forth like sky-rockets...and died almost as rapidly! This was real affection, real understanding, not mere carnal fantasy!

The tenor of his letter leads me to surmise that he had just finished a short leave-of-absence; the lovers had met, somehow, somewhere and parted, she to return home, he to rejoin his troops in the front-line. The same mysterious and discrete style, the same meditations and critical self-analyses. With a blunt m'Amie, he pours out his thoughts, just as they occur to him.

"Pale-faced" he writes, "with dark lines encircling his eyes, nervously crushing his hands together in a vain attempt to stop the tears which, in spite of all efforts on his part, will course down his dust-covered cheeks, a soldier of France walks along the road which leads to the front-lines of his sector. As he passes comrades, they turn to look at him, curiously...but refrain from asking embarrassing questions; there are so many private tragedies in war-time...one never knows!

"Ashamed of what he regards as a sign of weakness, the officer hastily, almost furtively, wipes his tear-stained face with a muttered "my God! how hot it is to-day" which, he hopes, will allay suspicions and stop the questioning which he dreads! Behind him, in the far distance, is the partly destroyed railroad station from the platform of which he had, but a few moments ago, watched a train, watched it as it slowly, haltingly and jerkily moved in a southerly direction, increasing it's speed and finally disappearing around a curve - carrying with it and on it his joy and happiness, his Amie!

"His poor brain is so befuddled that he cannot even recall having said good-bye to her; he must have remained there on the station platform, speechless, while she climbed aboard her compartment and the conductor slammed it's door; he had started to walk away before the whistle blew, then turned to see her...she had vanished and he began walking towards the trenches. Dear m'Amie, these are some of the thoughts which came to him as he walked: "Poor cynical philosopher, you had sworn not to allow yourself to suffer, you had steeled yourself not to hope for too much in this life; you had fully intended showing a complete restraint when the leave-taking moment arrived! How utterly you were mistaken in yourself; you who

thought you could control your emotions struck your flag and capitulated, surprised at finding yourself as ordinary men were, capable of suffering with the intensity of youth; youth, the age of idealism and of glorious illusions! You a cynic? You a philosopher? No! no!, for you suffer, you shed tears. A few moments back you were happy, now your happiness has left you, is moving further and further away. What remains?..... A sweetly mysterious voice, seemingly coming from nowhere, a voice which reminded me of your's, m'Amie, when you are in a very serious mood, gently murmured two words, Duty, Hope! The soldier awakened as from a dream: "Duty! They shall not pass; we must fight on to a glorious end for peace, peace and happiness for my valiant country! With redoubled energy he plunged into his duties and, as he fulfilled all of the tasks allotted to him he was sweetly encouraged and consoled by the feeling that Hope kept near him. M'amie! m'amie!, thou who art in truth hope, give me strength to do my duty to the end!"

Only a sincere lover could write as he did; he seemed to have had a passion for indulging in meditations and forwarding them to his dear friend.

Too overcome to fully answer her lover's gloomy outpourings, m'Amie apparently dashed off a short but passionate note to reassurance, of hope and confidence in a future which, at heart, she knows to be full of doubt.

"I am peering," she writes, "into the darkness each night, asking the clouds to part that I may see my own dearest Ami! Chéri! chéri!, there is nothing in the world that I would not do for you. There is no task that I would not cope with if to do it would help you in your life's ambition or make you happy. As the days draw nearer we shall look back and see a life of fullness, overflowing with love, kindness and devotion to each other. Would that you were here, but love will find a way and when it does it will find us ready. May you be happy, dearest 'till we can again be together and this lonely time will flit before the boundless wealth of joy, like the fleecy clouds soar before a mighty wind. Sweep me, dearest ami, into your arms and hold me, hold me forever! I am yours!"

Apparently dissatisfied with her short letter, she dashes off a longer effusion in which she tries to persuade her lover - and herself - that their future must end in complete victory!

"To-day, like many other days, I have thought and hoped and wished to be with you. At times it seems that I would fly through the air to get to you. I feel that I could

resist the natural forces of the earth and overcome any difficulty. I want to fly directly to your arms. Oh!, if I could only have you, chéri; life is of no avail unless I can have you forever and soon! With you, life is a pleasure. All the ordeals of life are easy to bear when I think of you and that some day I will be with you, never to be parted from you. Why must I suffer this unending agony of wanting to be near you when I feel that an unsurpassable obstacle is between us. It will pass away, it must! Love will endure. It is having it's extreme test. It will win out in the end! Somebody wrote in a poem that "love is more refined by sorrow" - "La douleur exhausse l'Amour" - then, chéri, ours is all surpassing. To be with you is an inspiration, a boost to keep up life's ambition. You are a source of ceaseless encouragement and never-ending help. You urge me on to useful accomplishment. I want to feel that I mean as much to you as you do to me, I shall always try to be a willing helper and, together dearest, we can overcome any difficulty. We shall climb together, hand in hand, and if ever a foot slips, the other will be there to catch and help over the rough spots and give a lift that will take him directly to his goal! M'ami!, I dreamt last night that you were here! Oh! what joy! Such a wonderful surprise! Would that you were and that I might know you were going to stay with me, for me alone! You encourage me to do things I wish I could do, and make me do them oh! so easily. Without your kind and patient help I would only live in hopes and probably die in despair. Chéri, only to know that you care, that you are standing by, urging me on with a nod, a smile or a kindly glance out of those kind, expressive brown eyes "tes yeux pleins de bonté, d'un brun si expressif" -, makes me tingle with energy that surges in every fiber of my being, whipping my effort to the utmost and giving up to nothing. No test could prove too difficult with you there. With you to urge me on I could overcome obstacles, in fact I would see no obstacles but would go directly to the goal. I should win or die in the attempt for I would never give up, knowing that my "petit chou chéri" approved! Before you came into my life I simply flitted to and fro without an objective, without purpose. With you life has a definate goal, life is worth while! Life is service and I mean to serve you, chéri. I want to do all in my power to be a source of joy and pleasure to you, I want to be and feel that I am just a wee bit "un tantinet tant soit peu" necessary in your life and that with me you too, can accomplish things that might not be as easy without me. You must tell me if I ever cease to help. When I get to the point when I am no longer of help, I want to die! When, oh! when can we be together? When, oh! when can I know you will not go away and leave me? Dearest, please don't leave me again. I don't want to be alone without you. May you never be as lonely as I have been! I suppose you have been very busy, in constant danger, always on the "qui-vive", always watching, watching, ready to shout "aux armes!"? I too tried to keep busy doing things that you might be proud of me but oh!, at times I've been so hopelessly lonely, so very, very heartsick! I am proud of you, chéri, of all you are doing, of all you have done and I shall always be proud to be your sweetheart and I shall attempt to do and live in such a way that you may be proud of me. Never once must we feel that life is worthless, but keep each other so desirous of doing for the other, that service will be joy. Some day when we are blessed, we shall sit at the fireside in the twilight of the day and, clasping hands, think of the sweetest hours we have spent and wish them multiplied for our own. Will you come soon? Can you come soon?"

It was this letter in particular which suggested to my mind the possibility that the two lovers were, in peace-times, writers and that they intended publishing their innermost thoughts at some future date.

"Emptiness surrounds and grips me, worse than ever! No longer am I conscious of Hope; it seems to me that I have lost it, forever! How I suffer! Oh God!, how I suffer!

"Dear m'Amie, I remembered the other day, a particularly black one for me, that there were bottles in my army locker....Cognac, Chartreuse and other liqueurs which friends had given me; they were within reach of my trembling hands, why not try to forget? I took my keys, lifted the lid and thrust my hand in, blindly, eagerly, expectantly; oh! m'Amie, it was your photograph which I grasped and let fall at my feet and reverently picked up again! A living vision of your dear self arose before me and murmured: "No, m'Ami, No!". I adoringly kissed your picture, replaced it and closed my locker; as I did so, the conscious feeling that Hope was floating over me returned, and I blessed you for giving me strength to resist a coward's temptation."

In another letter, in which a charming mixture of candor and doubt are blended, he says:

"I know what I sought and found in you; I have always known it, but this last separation more fully proves it. It was a soul which could answer my soul; it was the sweet and open interchange of intimate ideas and opinions, the joy of sharing in your joy and partaking of your gaiety and happiness! That is what I found, that is what I miss and long for as I write. Dear m'Amie, there are of course other souvenirs....equally tender and intimate....which mingle in my thoughts of you, but the dominant note which surmounts all others is the call of another soul who seems to understand me, nay, even seems to appreciate me a little!

"And you, m'Amie, there is the great question; have you unveiled yourself - "t'es-tu dévoilée" sufficiently so that I might really fully understand and appreciate you as you should be; have I been able to create in you a desire, be

it ever so small, to resume at some future date the dear talks, the precious community of spirit, the loyal and glorious companionship of the days gone by? Suddenly, as I write, I felt an intangible something floating around me, Oh Hope!, is it you, is it you?"

Eagerly did I seek and read the answer to these frankly masculine outpourings. Womanlike, m'Amie maintained a mystic, infinitely more poetic style; her reaction to the lover's candid hopes and beliefs is in the form of a confession, for her soul is evidently troubled with doubts.

"A strange being took me by the hand this evening and led me out into the black darkness. A starless night, uneven ground; at my feet were waves, angrily dashing themselves against the rocks; above and around me the wind whistled, cried and sighed. All about me the crackling and snapping of dead twigs....and I, alone in the night, followed this strange being who led me on and on, yet repelled me by his seeming indifference. In my agony I cried out: "I understand you, stranger, I know what you mean; you have removed me from all mortal contact, from all friendship, all beauty, and now you are going to say to me: Who are you, where are you going, what have you done with your life? You wished to know, to understand life and, suddenly, you understood...what was quite impossible! You wished for love; poor little girl, the ultimate joy of love is not for you. You believe that life's greatest task is to give happiness to those whom you love; are you certain that you have not planned their ruin, that you are not the cause of many a wrinkle, of many a whitened hair; can you be positive that you have not upset a life? Following nature's example, you gave yourself; in so doing, poor senseless girl, you commited your most grievous mistake, for nature has it's own laws...and they are not for you! That is why you accuse nature of having betrayed you and, in your anguish, call her cruel! You do not wish to repress the passionate feelings which burn within you, stir you up and, at times, give you the illusion of being a super-woman! You want these passionate flames, diabolical or divine, to mingle, sparkle and be eternally renewed; you wish to become greater, more loving and more beautiful than ever before but, for the first time, you now seem to understand why you stumble, question and rebel; why you almost envy the fate of X.....in his monastery! Daughter! daughter!, have you served God, even once, in your life?"

Judged from a purely litterary point-of-view, I have seldom read anything more sublime than the hitherto unpublished confession of this poor distressed soul! This unknown young girl reveals herself to be an intellectual of the highest type; she handles her pen with the touch of an artist and knows how to vibrate in unisson the entire gamut of human emotions. One senses that it was for love only and willingly, with open eyes that she had given herself to her lover - for love and, perhaps also, out of curiosity. Her musings suggest that, for some time past she had been propounding to herself the eternal question: "What is life?" Did she, at times, regret the step she had taken? A difficult question to affirm or deny; let us rather incline to the thought that she had brought happiness to a man who loved her, sincerely and honestly believing that in so doing she had obeyed nature - whom she adored and venerated - but that the social obstacles, delicately hinted at, which made a complete overflowing of her desires impossible, brought about introspective moods during which she, figuratively retired within herself in order to analyse, dissect her very soul, and indulge in peculiar auto-confessions. Just why it would have been indiscreet to name the monk in his monastery is a hard question to fathom; there must have been some mysterious reason for omitting it. Dates, places were always left out; the lover's censorship was most efficient.

In her next letter, the young woman discards the mysticism which characterizes most of her correspondence and fairly outrivals the French officer's frankness.

"I am alone in my little bedroom," she writes, "a room which, I am certain, looks upon me as almost a stranger; each piece of furniture, each chair, seems to have completely lost contact with me. There is, behind me, a tower of strength which constantly draws, links and binds me to you, m'Ami, a strength which follows me, inspires and helps.....yet, at times, causes me to become somewhat absent-minded and....nervous! During one of my restless nights, the anguish which I felt over the loss of that other self became so overwhelming that I sobbed and sobbed - God only knows for how long! The reaction was excellent; next morning I arose, dressed, looked about me with indifference and, thanks to my mirror, was able to converse with the past and tell it how happy and proud I felt that it had been.....what it had been! I regret nothing, m'Ami; I am proud to think that I yielded to your entreaties, proud to know that I belong to you, body and soul!

"Beloved one! when I think that, far away, far from me, in hourly danger, your wonderful soul is suffering tortures, there is no part of my being which does not call you, which does not thrill and yearn to stretch out towards you in order to appease the unhappiness which crushes you, smothers you and gives to your eyes that look of agony which I remember so well; if only I could, if only God would allow me to calm all those thoughts which lie somewhere in your brain, behind those expressive brown eyes which used to plunge directly through my very soul in order to read everything, even my most hidden thoughts! My helplessness overwhelms me; I open my arms to you, despairingly, alas!, they remain empty; I speak a beloved name; can he hear me? I go to my window and beseech the night; amid mysterious silence life continues it's

course; the stars twinkle, the moon shines brightly and the air is freezing cold; everything seems utterly indifferent to my sorrow!

Nature! nature!, is it your cruel wish that I should remain as I am now, inert and useless; is it your wish that, forever, I must crush the desire to give, to love, with which you so richly endowed me and by means of which I gave so much happiness to m'Ami? No! no!, it cannot be; you know it cannot be thus and, to-night, you have made me feel that more was to be expected of me; a little of your pity seems to have reached me - did you give it in anticipation of future trials, oh! Nature, did you? It matters not if nor how much I suffer, oh nature-god; only one thing do I fear; not to be your true disciple!

Two tears ran down my cheeks; I shivered under my black silk kimono and cast a last look at the dark night!"

As already stated, it seems quite evident that the unknown couple had decided on a certain style of writing, and that the French officer, after reading his Amie's letters, was in the habit of sending them back to her for safe-keeping. It is also quite possible that they may have intended publishing their innermost thoughts as soon as opportunity offered. Finally, one cannot quite dismiss the bare possibility that these charmingly artistic outpourings were merely produced in order to cheer the officer and help him bear the morbid tediousness of trench warfare. All over France, thousands of war-godmothers or "Marraines", as they were called, kept in touch with lonely soldiers or officers up in the front-line trenches, sending them "love packages", letters, and post-cards at regular intervals although, as a general rule, the two correspondents never met. Could it be possible that so much apparent sincerity was only intellectual flirtation? Surely not, and yet - one can never tell!

"Love is a flower of rainbow hue,
"Nurtured by heavenly light and air;
"And trust is but the deathless dew
"Keeping it's leaves forever fair!

M'Ami did not use this poem because he probably was not
familiar with the lines, but he must have had somewhat similar thoughts in his mind when he penned his dissertation on Love, in answer to her confession of doubt.

"Trust me," he writes, "for you must know by this time how profound is my affection for you. Often have I thought of the possibility of loosing you; we discussed that subject quite openly not once but many times. Should that ever occur, I would know that you have taken the decision which will give you the greatest permanent happiness; as for myself, I would continue to

bless you for the happiness you have given me, that you allowed my soul to meet your soul understandingly and that it was given to us to taste of such bliss as is seldom vouschafed human beings! You and I, m'Amie, have never hidden our thoughts from one another; we looked nature in the face and took our decision, freely, openly, fearlessly. Cursed be the apology of a man, "la négation d'un homme" who, by means of lies, false promises or threats, seduces a young girl or a woman; such a being is even lower in the scale of humanity than the libertine who, through sheer egotistic laziness, pays for his carnal satisfaction in order to avoid the trouble of flirting! No animal falls to such depths; courtship is a universal gift of animate nature! Mere physical desire is not love, only it's complement. True love represents an irresistible evolution which draws two beings together; if really lofty and pure in it's inception all obstacles will be overcome, no matter how impossible they may at first appear. It's supreme moment is the intellectual and physical blending of two perfectly harmonious, coordinated and mutually completing individuals. The different stages of this evolution are: mutual attraction, which can be considered as a tender psychologic and physical state of compatibility; trust, which cannot obtain in persons of opposite sexes unless both feel within themselves a strong desire, an urgent need and the possibility of telling each other the truth on all subjects and at all times and, finally, friendship, based on attraction and trust without which it cannot endure; friendship which, imperceptibly at first, finally leads two beings to a state of mind where they freely open the mirrors of their souls and exchange increasingly sweet, intimate and sacred confidences!

The sacredness of love! How truly does that word depict what I have in mind, m'Amie! Everything which the loved one prizes: reputation, honor, health and happiness, is and always must be sacred to the true lover. He feels the responsibility of a soul and tries to be worthy of the faith which his beloved one manifested in him when she placed her life, happiness and future in his hands! This is, indeed, a sweet responsibility which does him honor and which he, in turn, will always honor!"

Among the collection of letters are three, which depict the classical lovers' quarrel and reconciliation. Delicately hinted at, it started as the result of illness or physical exhaustion on his part. He fails to make her understand; she fails to believe him. Her letter is full of the bitterness of disillusion; his, full of solicitude regarding her future - both leave the door open to a possible and secretly hoped-for reconciliation. She evidently dashes her thoughts off while still under the keenest stress of intense and unsatisfied emotion.

"Well," she writes, "I played and lost, didn't I? Of course it wasn't a game while it was going on - I thought it was the real thing - but it amounts to that when one looks back at it. I only hope you think I was a loyal pal - "un copain loyal" - in the last interview; never could I disgrace my name by playing the weak coward! There was the "petit-bleu" - special delivery note, to be sure, by way of partial explanation and I knew, though you did not dream I did, what years of anguish passed over you the night you discovered - or thought you discovered - the death of exstacy for you, and I could think, as the silence lengthened, that just seeing me might be to you a poignant reminder of your deprivation. But if that were so, why, why couldn't you have forewarned me, instead of making me suffer the agonizing chagrin, rehearsed over and over in all it's bitterness, of feeling that perhaps you had wanted but one part of me, that all the talks on all sorts of subjects and activities which I had enjoyed so keenly because I thought the pleasure was mutual - all those talks were only to win one end! Oh!, I could cry out with the pain that stabs and gnaws "la douleur qui me poignarde et me ronge" so mercilessly whenever I think of it! It somehow never occurred to me that you would find no pleasure in me when the one thing was gone - you may say I am unjust, and I do try to crush the thought, for I hate it so for you and for me but, I cannot believe you! I'm not censoring you; it only seems to me that I have travelled through so much needless pain because you somehow didn't come and have one frank talk! One can endure a clean blow "un coup en plein", so much better than gruelling, incessant pain!

"You have taught me Man, but it is a knowledge dearly bought. I have looked all around the situation over and over, you may be sure, and in spite of the fact that you had thought it all out beforehand, it is quite possible that later thoughts came to you in regard to "legitimate relationships"; but you know that I wished to guard against any possible untoward happening in other's lives as jealously and completely as you; maybe more so since you taught me to modify my principle in regard to acts which might make others unhappy by the addition "tant qu'ils ne savant rien de notre liaison" as long as they do not know about our liaison! And, if you did change your thoughts or have others, you needed only to tell me.

"It is useless to say my love is dead; it isn't. Vital emotions of that sort and intensity can't be killed very easily, but I shall be able to stifle it very well in time. There can never be an experience with another. I told you our's would be, could be the only one when the supreme surrender was mutually made!

"I wonder if you are wishing this experience had never been, and are breathing sighs of relief "soupirs d'allégement" that it is past? It seemed to me so beautiful, so real, so vital. I could not have dreamed that so brief a life was possible to it, but I must say: "Volte-face, en avant, marche!" - about face, forward, march! Pardon me if this lengthy epistle should annoy you, but I could not refrain from giving you this added glimpse of "me" I laid bare so freely to you in the happy days!"

The "petit bleu" is not among the other letters but, in desperation, the soldier of France again tries to explain the raison-d'être of their misunderstanding.

"M'Amie," he writes, "words cannot describe the despair which overcame me when I failed you! Man is proud of his virility; the suddenness with which the physical change came over me crushed my vanity! I tried to tell you; you accused me of unfaithfulness. In my petit bleu note I swore to my innocence, assured you, on my word of honor that you were mistaken - all to no avail! What else can I do or say in order to make you believe me? I shall hope always that the happiness and satisfaction which is your due may come to you and remain with you. I cannot write any more!

So near and yet so far! What pawns we are after all in this game of life and love! Like other men, I cannot follow your moods or caprices and, unfortunately, I cannot take desciphering very patiently! I cannot say Adieu; it must be Au-Revoir!"

He quite evidently seeks m'Amie out on his next leave-of-absence, and the lovers resume their sweet intimacy.

"My heart has been so buoyant and swelling and smiling with happiness ever since the miracle of two days ago that it sings through my voice and smiles through my lips over and over. Oh! m'Ami, the wonderful, ever-perfectly vivid miracle of so long ago in man's time, but new as yesterday in reality, has cone flooding back and I live again and wonder again in it's mystery and ineffableness! Over and over I hear your repeated: "you know I do; you know I do", saying so much more than more explicit expressions from others - because you are you!

I am thinking you are expecting this word from me. Somehow my pen halts very often because, to-day, it seems difficult to say just what I would.

It is so wonderful to know it is all alive, just as ever! I am sweetly conscious of you, chéri, chéri! No more quarrels, no more misunderstandings but, promise me that you will not in future be so taciturn and inarticulate! I have tried at times to piece together what you have told me of your background and that I must remember that, when you seem different from others and hard to interpret! On my part, I promise to decipher my beloved enigma correctly, always and forever! I want you to know how joyously happy I am now, and why my thoughts and feelings will not reduce themselves to the compass of mere words - but I see that I keep on trying to make them! "Mon petit chou chéri, je t'aime" - I love you, my little cabbage!"

The gem of the entire collection is unquestionably the letter in which, for the first and only time, the young girl uses a heading.

"Our Anniversary!"

"Languidly seated on a large cushion, in front of the parlor chimney, hands clasped around my knees, I gaze at the warm flames as they play about the logs and.....I dream!

I see a larger chimney, within which larger, more impetuous flames escape, skywards; the room is darker, larger and colder; seated nearby are two beings who silently watch the logs as they burn and change to ashes.

Their faces, reddened by winter's cold, gradually relax; an infinite calm seems to permeate all things, yet the expression on their faces is one of tense anxiety and they avoid looking at each other.

She, full of nervous intuition, senses the struggle which is going on in her friend's mind; she would like to appease it but is afraid to move, lest she precipitate what she fears, yet wants! She would like to believe that the man beside her is happy, yet she knows and refuses to acknowledge to herself that she is the cause of his torment. She tries to consult God.....and ends by deciding that she will leave fate in his hands, so profound is her faith in him. She knows, she feels that she ought to flee from the room yet, as one paralysed, she sits......and waits!

He remains seated, gazing fixedly at the fire, a picture of intense muscular tension; his hands jerk spasmodically as though they wished to clasp something, yet dared not do so. The rising and falling flames play upon his face and seem to give it a stronger, more manly appearance.

She does not move, and hardly dares breathe.... presently, after a seeming eternity of waiting, her hands are clasped by two warm hands, adoringly tender in their virile strength....she hears, at a distance, the tumultuous beatings of his heart and now, at last, listens to his confession of love, his halting words and unfinished sentences. Slowly, very slowly, the maiden lifts her head up to look at him...and sees two love-tormented, supplicating eyes, in whose dark depths shine a flame the like of which she has never seen before! Her enraptured gaze seems to enter his very soul; his heart beats are now closer, more rapid;.....she closes her eyes and, as a flower too heavy for it's stem, her head slowly rests on the shoulder of the man she feels herself loved by and she opens her lips to his in order that he may gather her soul unto his in the first kiss!"

The German avalanche of March 1918 brought this unique and intriguing correspondence to an abrupt end. Were the lovers ever able to re-write their former thoughts; did the French officer survive the terrific battles which succeeded each other until the Armistice - or does he lie under one of the hundreds of thousands of modest tombstones erected in France and inscribed "La Patrie reconnaissante", the grateful country?

I hope, let us all hope, that he did survive, that they met again and again, that all their difficulties were swept away and that to-day, somewhere in France, these two rare souls now have the right to openly proclaim what they wrote to each other during the World War!

George Skitarnow.

A Corner of the Great Battlefield

An American Surgeon in the Midst of Things

March 21 – April 3, 1918

Without mentioning names of places, a short description of our location will explain the immense importance of this hospital during the battle. The first great rush brought the battle line seven miles north of us. We occupied almost the apex of a triangle, the base of which represented the battle line, the two sides with two converging valleys along in which the Boche hoped to push on to Paris.

FRANCE, April 12, 1918.

To the Editor: The great battle which began in the early morning hours of March 21, '18, found the Ambulance des Alliés with some two hundred wounded, seventy-five of whom were American, one of our divisions occupying a nearby sec- tor jointly with the French.

In order to have all available beds possible at our disposal, the latter were evacuated on Paris by train and from thence by ambulance to U.S. Base Hospital at Neuilly. The French wounded were sent to the interior by train. All stood the journey very well. A few of the patients recently operated on were kept for a subsequent evacuation.

An atmosphere of tense excitement and uncertainty surrounded us. We knew that the thinly held British line had been overwhelmed by sheer weight of numbers and that the Tommies were falling back, making the enemy pay dearly for his gains. Just where the retreat would end was anybody's guess. The highway in front of the hospital was a seething mass of soldiers, artillery, munitions, etc., speeding north to check the invasion; an equally dense, infinitely pathetic mass of civilians were trekking south. Baby car-riages, wheelbarrows, two and four-wheeled carts, and car-riages of all sorts passed before our eyes in an apparently unending stream. These poor people had just begun to live their lives over again, amidst familiar surroundings, when the stern necessities of the battle forced them, for the second time in four years, to abandon their homes and fields in order to seek shelter elsewhere.

Two girls, both under 20, stop in front of the operating room windows and lean against the parapet to rest. Taking turns, they had trundled a wheelbarrow, on which was a single black trunk, for miles along the dusty road. They, at least, will not fall in the hands of the Boche! An old grand- mother pushes a baby carriage in which a pretty 4-year-old girl sits, clutching a wooden doll in her hands, while at her feet is a wicker basket containing a hen and it's chicks! In a two-wheeled cart lies an infirm grandfather, with a goat and calf for company, while his daughter walks beside the horse's head and three boys tramp behind, clinging to ropes which dangle from the vehicle! The pictures are pathetically funny, and one tries to laugh - in order not to cry - for men have learned to cry in this war; cursing is inadequate to express the feelings which surge within!

Wounded, both English and French, began arriving on the morning of March 24; the earlier signs of military confusion had ceased, and the French had jumped into the breech, while the English, worn out by the terrific onslaught, but still full of fighting spirit, fell back on the town of ...... to reform. It was a magnificent exhibition of team work, carried out in the very midst of a desperate battle for the road to Paris! Three divisions against eleven, and the eleven could not get through!

LOCATION OF THE HOSPITAL

Without mentioning names of places, a short description of our location will explain the immense importance of this hospital during the battle. The first great rush brought the battle line seven miles north of us. We occupied almost the apex of a triangle, the base of which represented the battle line, the two sides [with] two converging valleys along in which the Boche hoped to push on to Paris. Had he been able to pour down either valley, our retreat toward Paris would have been cut off, as the junction of the two rivers takes place three miles south of the hospital grounds.

With no railway lines, bridges, or canal locks within a mile, and no munition depots within three miles of us, we enjoyed a reasonable degree of security from shell fire or aerial bombs. The tiny village did, however, shelter troops who were constantly billeted there, and a long-range battery installed three hundred yards from the hospital barracks put us in real danger. The guns, most fortunately for us, only remained twenty-four hours, departing before the Boche avions had a chance to photograph their location. Shell fragments, bullets, and even pieces of shell casings from our own antiaircraft guns sprinkle the roofs and grounds about us, but this small diversion has been of almost nightly occurrence for the past month, and we have grown accustomed to the avion raids. The vicissitudes of the great battle thus made us the nearest surgical formation in this sector, capable of operating on all "nontransportable" cases. Owing to their exposed position, all the large hospitals east and west of us had been compelled to evacuate on the second day of the battle.

The fact that only two surgical teams, one French, one American, were attached to the hospital when the attack began, was a serious handicap, but the American Red Cross promptly sent reinforcements, with a supply of food and four motor trucks, the latter placed at our constant disposal in case a hasty evacuation became imperative.

By Monday, March 25, the war news was so conflicting, yet so generally discouraging, that the directress of the hospital very suddenly ordered a general evacuation. At 2 o'clock, on the morning of March 26, everybody was awakened and told to be dressed in ten minutes! The peculiar arrangement and combination of circumstances, which allowed a civilian directress to order a military evacuation, are not germane to this letter, and will, therefore, not be mentioned. Suffice it to say that the French staff and personnel obeyed said order to the letter.

Approximately one hundred and fifty patients were transferred by ambulance to the hospital boat belonging to this formation, with the French staff and personnel and all of the nurses, the evacuation consuming four hours. As the boat was docked in the canal, a mile from the hospital, the transfer was extremely rapid and easy. Needless to say, there was no opportunity for cleaning up the wards or even storing away any of the material. Chaos and confusion were left behind.

THE HOSPITAL MOVED - NOT

Major John J. Moorhead, M.R.C., U.S.A., and myself, not being under the direct orders of either the directress or the French Service de Santé, expressed our firm desire and intention of remaining behind. There was plenty of work ahead of us; the Chief Surgeon A. R. C., under whose orders we were detailed for temporary duty, had not ordered us to evacuate, and we both felt that we could make a quick get-away if the military situation absolutely demanded our taking such a step.

Our request was granted. One patient, suffering from a lesion of the left femoral vessels and pelvic peritonitis, was left behind. By 6 a. m., March 26, the staff and personnel of the hospital consisted of Major George de Tarnowsky, M.R.C., U.S.A.; Major John J. Moorhead, M.R.C., U.S.A; Corporal Bonne, French Service de Santé; Corporal Brandinière, cook; Mme. Louis, kitchen helper; Louis, blind retainer; and Jean Legros, gardener.

The two French surgeons envy us our good luck; there is, however, plenty of work for them to do on the boat, as many of the wounded are evacuated before operation. The nurses, English, Irish, Canadian, and American women, without a single exception, volunteered to remain with us; we unfortunately had no military authority over them, and they were compelled to go with the French staff.

Mme. Louis, fearful lest her blind husband be left behind if we have to evacuate, makes me repeatedly promise that he will be put in one of the A.R.C. trucks. She, on the other hand, swears that she will put her chickens in a crate so that we can have fresh fowl to eat all the way to Paris! Neither husband nor chickens must be allowed to fall in the hands of the Boche!

Left to our own resources, Major Moorhead, Corporal Bonne, and I prepared forty beds, sterilized several sets of instruments, and stored away the uniforms, arms, etc., which had been left behind. We fortunately have a large supply of sterile dressings, cotton, and bandages.

Using the A.R.C. supplies which had been left under our charge, we distributed corned beef, cheese, crackers, chocolate, etc., to scores of hungry soldiers who passed the hospital gates. The cook presided over this distribution with much gusto, adding coffee or wine from the hospital stores.

Wounded soldiers began to arrive early in the afternoon, and we prepared to operate on the urgent ones. At 6 p.m., Capt. Peter M. Keating, M.R.C., U.S.A., very opportunely arrived from Paris and was immediately impressed. The electric plant having been disconnected, candles, stuck on a wooden board, and two lamps served for lighting purposes. This compels us to use chloroform instead of ether anesthesia.

From 7 p.m. to 4 a.m., we three surgeons performed twenty-six consecutive operations. The A.R.C. truck drivers were impressed as stretcher bearers and orderlies, while Corporal Bonne attended to the sterilizing. At 10 p.m., several American volunteer nurses, belonging to what is known as Mrs. Daly's équipe or team, arrived from Compiègne and immediately rendered invaluable assistance.

We had barely retired for a few hours' sleep when at dawn of March 27, several A.R.C. officials drove up with very alarming news, advising us to evacuate immediately! All of the cases operated on were loaded in ambulances and trucks and taken to Compiègne, five miles south of us, where they left on the last sanitary train, all the hospitals in that town having received orders to close up and depart.

Before coming to any decision regarding our small, there-fore very mobile, staff, Major Moorhead and the writer drove to the hospital center of Villers-s/Coudon, three miles northwest of the Ambulance des Alliés. The C.O. informed us that he had already received three evacuation orders and as many counter-orders, the last one to the effect that he must evacuate all wounded, but that his hospital would continue to receive slightly wounded cases; the latter, however, were not to be operated on but merely dressed and put on a sanitary train. We were assured that any further evacuation orders would be transmitted to us by courier, as our phone had been cut; this promise was most welcome and reassuring.

After holding a consultation with Major Moorhead, we decided to stay. Late in the afternoon, our French C.O., with the two French surgeons, returned from the boat which they had left below the Compiègne bridges. With fourteen volunteer American nurses and two surgical teams, we were able to operate steadily throughout the day and night.

At 3 a.m., March 28, a courier from Villers-s/Coudun Hospital arrived, with a message from the C.O. to the effect that he had just received orders to evacuate all wounded and prepare to retire with his supplies and personnel. While our French C.O. went to army headquarters nearby for final orders, the wounded were prepared for evacuation and sent to the boat which had docked in the canal, eight miles south of us. Returning from headquarters, our C.O. reported that we would be allowed to remain at our post for the time being, but that there should be a daily evacuation of all wounded after their operation. Our canal hospital boat, propelled by it's own steam and fitted up with seventy bunks and a small operating room, had been in commission about six months, rendering good service. It's utility, during the battle, was considerable. For instance, on March 24, in the thickest of the fight, it went up the canal to Noyon and came back with 180 wounded, seventy of whom were "lying" cases, the balance being able to sit here, there and everywhere, above and below decks. During two of the "en masse" evacuations from the hospital, forty-six operations were performed on board the boat by the French surgical team.

March 28 and 29, the operations continued steadily, one surgeon at a time managing to get a few hours rest. A second American team, consisting of Capts. Charles Farmer, Philip Wilson and First Lieut. Thomas S. Hardy, all M.C., U.S.R., arrived from Paris. Unfortunately, Captain Keating was recalled to Paris so that, during the absence on the boat of the French team, we were still reduced to two operating teams, with a spare surgeon for stimulations, dressings, etc. On this same date, March 29, seven of our nurses returned, and the Mrs. Daly équipe departed. While new wounded continued to tax our operative capacity to it's utmost, the general news of the battle became more reassuring; the Hun flood had been brilliantly dammed. We began to keep our operated cases until all danger of secondary hemorrhage or sepsis had passed, only evacuating when our nursing capacity was absolutely overtaxed.

Owing to our nearness to the firing line, the entire nursing staff, for the boat and hospital, was kept down to twelve nurses and their matron, with some thirty French orderlies. This personnel enabled us to hospitalize as many as seventy-five severely wounded cases at a time, but no more.

By Wednesday, April 3, we had definitely organized three operating teams, working turn and turnabout in ten hour shifts. A ten-hour shift was decided on in order to give each team an equal number of night hours in the operating room, admissions being heaviest after nightfall. When not operating, we received, diagnosed, and prepared cases, dressed our own cases and rested whenever we could.

Early in the battle, our trunks and lockers had been sent to Paris for safe keeping and, with the contents of a suitcase for our entire wardrobe, we lived from day to day, stripped for the surgical fight which it is our lot to wage. For almost two weeks, we had no postal service; at present, we occasionally receive mail and can write about the weather and the safety of our persons to the loved ones at home.

Distractions abound; not a day passes without several aerial fights; bombardments start up at all hours of the day and night, and we keep wondering whether Fritz will not drop a shell or bomb "by mistake" on our modest Red Cross group of buildings!

While Major Moorhead and I have tabulated all of the cases which have passed through our hands, there has been no opportunity to classify the operations and results. Suffice it to say, for the present, that during the first great rush, i.e., from March 25 to April 3, when the three teams began their regular schedule, 149 operations were performed, with a total mortality of six cases. Two of the six were moribund on admission, no operation being possible; of the remaining four, two were intrathoracic and two intra-abdominal wounds with profuse hemorrhage. One hundred and three operations were performed at the hospital; forty-six on the boat.

Just how long this emergency is to continue, nobody knows. We intend sticking as long as there are wounded to succor and our orders remain unchanged. That we shall still be operating in the same hospital when this letter reaches you, is the earnest wish of

GEORGE DE TARNOWSKY, Major, M.R.C., U.S. Army.
Chief Surgeon, Ambulance des Alliés.

Evacuation Under Fire

The Redemption of a Society Hospital

March – April 1918

At two A.M., March 25th .... but let the formal report be set aside while the human side of the story unfolds itself!

Without any preliminary warning and - apparently - acting under the impulse of a sudden, overwhelming fear, the Directress of the Ambulance des Alliés ordered an immediate evacuation of the entire hospital! Surgeons, nurses, and orderlies were told to be ready in fifteen minutes: ambulance drivers were ordered to load all wounded and retreat towards the Paris zone.

Evacuation under fire.

by

George Skitarnow

--

Perhaps a better title to this incident would have been "The redemption of a Society Hospital", because temporary removal to a safer zone of the social element of the Ambulance des Alliés enabled this hospital to receive and treat in a few hectic weeks a larger number of wounded than it had hitherto hospitalized in a year and to become, during the emergency brought about by the German advance against the British 5th Army, the only hospital in it's sector capable of operating upon severely wounded soldiers. The Official report of this period is thus tersely recounted:

♦♦♦♦♦

Ambulance des Alliés
Postal Sector 164, France
May 4th 1918.

From: Major G...T... Chief Surgeon
To : Chief Surgeon A.E.F.
Subject: Report on the Ambulance des Alliés at Annel, Oise, during the battles of March-April 1918.

1. While the battle itself began in the early morning hours of March 21st, it was not until the 23rd that the British retreat affected the normal status of the Ambulance. On that date, with seventy-five wounded Americans (26th Division) and an equal number of French soldiers in the hospital, the situation became rather an alarming one; signs of military uneasiness and confusion were manifest. An immediate evacuation of all transportable cases was decided upon, the French being sent to the Compiègne régulatrice, the Americans to American Red Cross Military Hospital No

1. at Neuilly. Ten Americans whose wounds were not yet in a satisfactory condition, were kept behind with about twenty French comrades.

2. The reports of the battle, although very conflicting, were not satisfactory, the English being manifestly hard pressed and vastly outnumbered. Many hungry and tired British soldiers and officers who hadlost contact with their units strayed into the hospital grounds for food and a few hours rest. The city of Compiègne, badly dammaged by aerial bombing and heavy artillery fire, was evacuated with all of it's tributary hospitals on March 23-24. This left the Ambulance des Alliés to act as the only advanced surgical formation, in a very wide sector, capable of treating "non-transportable" cases.

3. On March 24th our hospital boat was sent up the Oise canal to Noyon where it barely escaped capture by the Germans whose advanced elements were entering the north end of the city as the English rear-guard retreated down the Oise valley. It returned that same evening with one hundred and eighty wounded officers and soldiers. Twenty-six severe cases also arrived by ambulance. Operations were performed uninterruptedly for thirty hours by the American surgeons; the French surgeons continued the work for another six hours. All operated cases were evacuated on Paris; the wards were kept filled by constant new arrivals. Three motor trucks were placed at our constant disposal for evacuation purposes.

4. At two A.M., March 25th....... but let the formal report be set aside while the human side of the story unfolds itself!

♦♦♦♦♦

Without any preliminary warning and - apparently - acting under the impulse of a sudden, overwhelming fear, the Directress of the Ambulance des Alliés ordered an immediate evacuation of the entire hospital! Surgeons, nurses, and orderlies were told to be ready in fifteen minutes: ambulance drivers were ordered to load all wounded and retreat towards the Paris zone. To the north of us, the roar of cannon, wrang of air-bombs and flare of night fuses illuminating the battle-line from the Lassigny plateau to Mount Renaud, running eastward from the captured city of Noyon. On the "perron" or front steps of the chateau which constituted the main building of the hospital, stood the commanding general of our old sector and his bride of a few days - the Directress of the hospital! Both showed considerable agitation, the general frequently admonishing the tired and bewildered old French orderlies - men past military age or fathers of large families, called "pépéres" by the younger soldiers - to: "hurry! hurry!, or the Boches will get you!" Ambulance after ambulance, loaded with wounded, rattled down the hill leading from the chateau grounds to the Oise canal where our hospital boat lay, tied to the wharf. Nurses and servants hastily gathered their most valued possessions and a change of under-garments and gathered in the courtyard, awaiting their turn to be evacuated. Only two American surgeons were on duty at that time, the others having been recalled to their units or to other formations where their services were urgently needed. The shortage of Medical personnel throughout the duration of the War was one of the gravest problems our Chief Surgeon had to face and solve, hence American Medical officers were constantly being shifted from one duty to another.

Major J.J.M...... of New-York City and the writer walked around and around the courtyard circle trying to decide what to do. They had been operating or dressing wounds from eight o'clock in the morning until eight in the evening on the previous day - a total of twenty-six operations and as many re-dressings - and had retired to rest, leaving a French operating team to continue with the unending work, when the hurried summons awakened them from their well-earned sleep. General war news that afternoon had not been particularly bad, although the Allied retreat was still in progress at some points. American Red Cross officials had driven up from Paris in the late afternoon and placed at their disposal three trucks with the necessary drivers and supply of gazoline, in order to provide the surgeons with a rapid means of retreat, should this become necessary. The gazoline drums had been hidden under trees, three hundred yards from the nearest hospital barracks, in order to save the latter in case of fire from an aerial bomb or heavy artilleryshell. As far as one could judge, normal battle conditions prevailed. The retreat of civilians had ceased, no more English or French stragglers had been fed or treated during the previous day, and French troops and supply trains constantly passed in front of the operating rooms, trekking northward to stop the enemy. American ambulances had kept on bringing in wounded men, and none of the drivers had mentioned any change in their "loading points". A loading point is that area nearest the battle-line to which ambulances dare go, wounded being carried on litters from the spot where they fell to the first-aid station. The loading point is either at the aid-station or somewhat further to the rear, depending on the amount of protection from shell-fire which can be obtained. One patient, suffering from peritonitis, had been at first judged to be in too critical a condition to bear transportationand, as is customary in warfare, the commanding officer had designated the junior French surgeon to remain behind with him. A hurried re-examination caused the latter to change his opinion and the last of the two hundred odd wounded departed for the hospital boat. The entire French staff was, of course, obliged to obey the order of evacuation. We two Americans were, technically, in a somewhat uncertain position as regards our military status. [As] Army surgeons, we were on detached duty with the American Red Cross. This organization, anxious to help out the French army, had assigned us to the Ambulance des Alliés for temporary duty. Under the circumstances, and believing that there was plenty of surgical work ahead of us if we remained at our post, we decided to ask the French general if he would allow us to remain behind with such help as could be spared and would volunteer. Our somewhat unusual and quite unmilitary request was hesitatingly granted "at your own risk and peril", as the General sternly put it! At once, we mingled with the nurses, calling for volunteers; one and all, they enthousiastically offered to remain with us, but the Directress was adamant; go they must, and did! As the general and his wife disappeared from sight in the former's staff car, Major M. and the writer silently shook hands in the dark and looked around us. It was then six in the morning; the entire hospital staff was now reduced to two American surgeons, corporal Bonne, a middle-aged bearded Frenchman who in private life was a well-known Catholic priest, Brandinière [the] military cook, Louis an old blind retainer [and] his wife who helped in the kitchen, and Jean Legros [the] gardener. All of these brave people had offered to remain behind and help to the best of their various abilities. Telephones had been disconnected, electricity was shut off, and we had to do the best we could by candle-light. Chaos, confusion, and filth confronted us everywhere, the inevitable result of a panicky departure. Clothes, arms, equipment of all sorts, were hastily removed from the wards, stored away and, with the help of a stray boy from the village of Annel and an additional orderly who unexpectedly returned from his furlough in ignorance of the evacuation, forty beds were made ready. The operating room was our next care. We took stock of available surgical supplies and instruments, cleaned up floor, walls, and tables, and sterilized several sets of instruments. At irregular intervals throughout the day, hungry soldiers, both English and French, who had lost contact with their units drifted in the hospital grounds and were directed to the kitchen. Using the A.R.C. stores, Brandinière the cook, fed them on corned-beef, cheese, crackers, wine, and coffee and then sent them on their way. None of these soldiers being wounded, we were actually violating Red-Cross conventions but... we took a chance! As for ourselves, we snatched a bite whenever we could. At noon of the same day, Mrs. Daly, directress of a volunteer nursing unit known by the French as the "Équipe Daly", drove up from the city of Compiègne for food and surgical supplies she was in dire need of. We were fortunately able to fill her Ford truck with whatever she asked for. Her group of nurses, attached to Auto-Chir No 7 (a mobile hospital unit), had been bombed out of the town of Ressons-sur-Matz on our left, the night before. In Compiègne, the Daly unit had joined forces with the A.R.C. outfit and were kept busy dressing minor injuries and doing cantine work on the bridge crossing the Oise river. Few incidents endeared the A.R.C. to the French army and people more than did this cantine work. Imagine a pile of cases of corned-beef, cheese and crackers, huge urns for coffee and cocoa, and cartons of cigarettes. Opening the cases, a famous New-York Clergyman; around him devoted young American women handing out food, a cup of coffee or cocoa, and two cigarettes to each Tommy or soldier of France on his way to the battle-field - perhaps to death -, with the simple phrase “Avec les compliments de la Croix Rouge Américaine”- with the compliments of the A.R.C.! A simple gesture, but of what immense psychological strength! In front of the main A.R.C. cantine in Compiègne, an English general sat silent in his automobile for some minutes, watched the lines at the kitchens and then exclaimed: “Wonderful effect, a little thing like that! The men come in droopy, and they go off with their tails in the air!” Night and day, until the emergency died down, did these devoted Red Cross workers stand on that bridge, dealing out food for body and cheer for soul, sending brave men to battle with a smile on their faces and cigarette smoke halos drifting around them! French officers would stop their mounts and gaze upon this scene with tears of gratitude in their eyes, ride up to the young women, kiss their hands with a murmured 'Dieu vous bénisse, Mademoiselle”, God bless you, Miss, and then dash off at a gallop to rejoin their units. War may be as horrible, as futile, as useless or as stupid as some of our citizens proclaim it to be but, it also is full of sublime altruism, spirit of sacrifice, and comradship! We who have been through it know and feel this to be a fact!

Back at the hospital, wounded soldiers were brought in throughout the afternoon, their numbers increasing after nightfall. Groups of exhausted fighters drifted in at irregular intervals with conflicting accounts of the battle and it's outcome. Providentially Captain K.. arrived in the evening from Paris and was at once impressed into service. The three surgeons performed twenty-two consecutive operations in eight hours. A more weird operating room than ours could hardly be imagined: it was worthy of comparison with scenes from Dante's Inferno! Six truck drivers were given a candle apiece and made to stand around the operating table, shifting the flickering lights at the surgeon's command. Ether being explosive, only chloroform anesthesia could be used. Other drivers acted as orderlies, wheeling patients from ward to operating room and back again while faithful Corporal Bonne attended to the sterilizing of instruments and dressings and mopped up the floor between operations. All windows had been heavily darkened by tacking several thicknesses of blankets over them - lights at night invite aerial bombing - so that the heat soon became stifling. In the middle of the night, Mrs. Daly and four of her nurses drove up and relieved the tired orderlies until four in the morning when the exhausted surgeons retired for a few hours of rest.

Early in the battle, our lockers, bedding rolls, and duffle-bags had been sent to Paris for safe-keeping and, with the contents of a musette for our entire wardrobe we lived from day to day, stripped for the surgical fight in which it was our lot to wage. Each officer, volunteer nurse, driver, or orderly attended to his or her laundry - it was not possible to reserve the classical Monday for that important duty! Astonishing, how easily we managed for eight weeks with one change of under-clothing and socks and an extra campaign shirt! For two weeks, we had absolutely no postal service. What few short letters we managed to send back home had to omit all references to locality or battle conditions; chiefly, we wrote about the weather and tried to assure our families of the - to date - safety of our persons! Distractions consisted in watching "dog-fights" between English and German or French and German airmen: several wounded aviators were operated upon, and many a brave airman was buried with military honors in the little hospital graveyard. Bombardments kept one constantly on the alert, and a climb to the top of Gannelong hill back of the hospital, gave us "ring-side" seats for the almost daily capture and recapture of Mount Renaud. It was a thrilling sight to see a gas attack, launched in the Oise valley below, slowly drifting up the mountain side followed, a few minutes later, by waves of French soldiers. It was also thrilling, but sad, to see similar waves of gas fumes launched from the crest of Mount Renaud and, with a favorable wind, drift down into the valley, temporarily disloging the brave, stubborn Poilus! The burning one day of a French "sausage" or stationary observation baloon was an example of fiendishly clever air-tactics on the part of the German air-corps. The "sausage", stationed on our left near the village of Villers-sur-Coudun, was clearly in view. On that particular afternoon, just above it and running in a north-easterly direction, floated a greyish-white cloud, almost immobile in a windless atmosphere. Suddenly, from the south tip of the cloud, emerged three enemy planes, flying close together but at different altitudes. All three dove down, opened fire on the helpless baloon with incendiary bullets, did a half-loop with superb military precision and, almost before we could gasp or cry out, disappeared once more behind the protecting cloud! As the "sausage" burst in flames and began to curl up, we saw the observer climb out of the observation basket and jump. The parachute did not open, and it was found, a few minutes later, that the Frenchman had been severely wounded before he jumped. Another brave hero added to the long list!

Early in the morning of March 27th, several officials of the A.R.C. drove up from Compiègne with rather alarming war news and urged us to evacuate the hospital. We sent all transportable wounded to the Compiègne régulatrice - or railway regulating station - where they caught the last hospital train for Creil. Before coming to any decision regarding our small and therefore very mobile staff, we drove to the French hospital for slightly wounded at Villers-sur- Coudun and interviewed the C.O. His telephone had not been cut off as had ours so that he had been able to keep in better touch with hourly events. From him, we received the information that three evacuation orders and as many counter-orders had been received at his hospital in the past two days, the last one being to continue taking in slightly wounded and evacuating them to the rear on the same day. The C.O. assured us that any further orders would be promptly transmitted to us by runner. In spite of the uncertainty of the situation, we decided to remain. Well, it was that we did so for a portion of the French medical personnel of our hospital returned that same afternoon, after an absence of thirty-five hours! That same night, Compiègne being under heavy bombardment, all of Mrs. Daly's nurses took refuge with us.

At three o'clock in the morning of March 28th, a runner came from the Villers-sur-Coudun hospital with a note written by the C.O. stating that he had just received orders to evacuate all wounded at once and prepare to retire towards Creil with his supplies, staff, and personnel. The nurses and A.R.C. drivers were aroused and told to dress at once while a runner was sent to Corps H.Q. for final instructions. As a measure of precaution, we sent all the wounded down to the hospital boat. General Pellé's H.Q., which had been established on our right in the village of Ribécourt during the battles of Le Plémont and of the Lassigny plateau, had been subjected to such an intense bombardement that, during the night of March 27-28, it had been obliged to move southward to the village of Clairoix, some three miles behind us. Our runner was promptly challenged by a sentinel and directed to the latter village. H.Q. reported that the line had broken during the night but been re-established through a counter-attack; the situation was slightly better, and we might remain until further orders.

The efficiency of the German spy system was clearly demonstrated during that memorable night. General Pellé and his staff arrived at Clairoix at one in the morning of March 28th; exactly fifteen minutes later, German shells poured into the tiny village, compelling everybody to take refuge in the deep cellars in which the inhabitants had built in the lee of the Gannelong hill. Taking his habitual stroll through the wooded portions of the hill that same afternoon, one of the American surgeons almost walked over the sleeping body of a British Colonel, [the] liaison officer with French Corps H.Q.! With a profusion of vitriolic language, the exhausted officer detailed the events of the previous night!

By April third, the military situation had become well stabilized. Mount Renaud continued to change hands quite frequently, but the rest of the line held firmly. The Compiègne bridge having been heavily mined, ready to be blown up should a further retreat be deemed necessary, our hospital boat was ordered to remain below it, nearer Creil, all wounded having to take the ten mile trip to the dock in ambulances. On April 10th, General Pellé and his A.D.C. rode up to our hospital, thanked the American surgeons for sticking to their posts, and invited the Chief Surgeon to lunch with him at H.Q. the following day. He had moved back to a large villa in Margny-les-Compiègne, which had escaped the bombardment and, while showing traces of the terrific ordeal he had gone through, was very optimistic as to the final issue of the struggle. The relief of tension after twenty days of almost superhuman efforts to do all one could for the alleviation of human suffering and preservation of life and limb, the sincere, well-chosen words of thanks, the comradry shown a non-combatant by combatant staff-officers, made a never-to-be-forgotten impression on the American surgeon. Subsequent military decorations, while received with pride, could not efface such really personal marks of appreciation and gratitude as were showered on him that day. Though he knew it not at the time, he was never to see the general again in the flesh, for the latter died a few years after the end of the war. Among the surgeon's war treasures, however, is a beautiful letter of reiterated thanks, written from Constantinople where Pellé had been sent by France as it's High Commissioner. A brave soldier, ardent patriot, and fine gentleman-of-France was he!

A well stabilized situation merely meant that we could with reasonable certainty count on getting from four to six hours' sleep in the twenty-fours, and that threatened retreats were no longer looked upon as every day possibilities. Three operating shifts of ten hours each were organized, thus enabling each team to have it's full share of night work, which continued to be the heaviest. This system continued in force until May 27th, 1918 when the great "surprise advance" of the Germans from the Chemin des Dames to Chateau-Thierry completely outflanked us. The Ambulance des Alliés ceased to exist from that day on. The French unit reopened another hospital at Pont-Sainte-Maxence. The American surgeons were recalled to the A.E.F., two of them being put in command of hastily improvized American Red Cross Military Hospitals to meet the very urgent needs of our army. A new chapter of experiences was to begin - our real entry into the World War as American units. The first German offensive of 1918, a section of which it had been our good luck to witness was a great show - while it lasted!

In retrospect, every surgeon who had the priviledge of serving under our Allies, felt that, when the A.E.F. needed his services, he did better work, showed finer judgement, and probably saved more lives and limbs than did those among his colleagues who, arriving overseas late in 1918, were at once brought in intimate contact with a highly specialized problem - the proper treatment of war wounds!

In glancing over his war diary, the writer came across the following lines, dashed off - quite appropriately though unintentionally - on April fool's day 1918!

Evacuation!

--

This hospital's evacuated, that is to say,
It had evacuated, just the other day;
But this hospital's open, that is to say,
It was, I thought, quite open, just the other day;
And ere this plagiarised poem may appear,
More wounded heroes may be, have been, will be here!

Reference has been made elsewhere concerning the yeoman work performed on the Compiègne bridge by a New-York Clergyman. His report of Red Cross activities in that sector appeared in the November 1918 number of the Red Cross Magazine. My readers - if any - will, I trust, pardon the apparent but quite unintentional egotism which prompts quoting the two concluding paragraphs of what the good Domine calls "A typical Red Cross Report". After describing the varied activities of his co-workers, he goes on to say: "When a complete story of the prolonged battle shall be written, there will be few incidents which will stir Americans with a more justified satisfaction than the conduct of two American doctors at Annel, an outpost of Compiègne, on the night of March 25th, 1918. The patients of their hospital, under orders from the French Service de Santé (i.e.Medical Corps), had been loaded into canal boats and pulled away the preceding night. The French surgeons and all the nurses were gone. The Germans were within a half dozen kilometers and advancing with apparently irresistible momentum. Everyone felt that capture was a matter of hours. Heavy artillery, set in the hospital grounds, made a terrific din. But the American ambulance sections, long in the habit of bringing their wounded to Annel Chateau, kept on coming. They did not know where else to go. Two Red Cross doctors (i.e. Army surgeons on detached duty with the A.R.C.), Major T... and Major M..., quietly said that as long as American boys brought them blessés - wounded -, they would find American doctors to operate on them. In the deserted chateau, the doctors stayed on and kept on operating. In the nick of time, five Red Cross trucks with supplies arrived. From Compiègne volunteered two canteen women. That made the hospital personnel - the two surgeons, the two women, and the ten drivers.

Get the picture of that night at Annel. It represents America at it's best. It contains the very essence of what Red Cross spirit can be - devotion, resourcefulness, persistence, and a passion to aid suffering that outranks other emotions. Outside - the moonlight, the hum of planes, the shaking of the earth, the belching of sound from unsuspected places, the steady low detonation of the battle six kilometers away. Inside - the two doctors, not looking up from their delicate task, the canteen women handing them instruments and dressings, the Red Cross truck drivers holding candles, bringing in and carrying out the wounded, and then playing hospital angels, in an unknown tongue, to the fellows lying in the wards. A day and a night and a second day, the two doctors operated. The crisis passed. The regular attendants returned, and routine was resumed. The doctors are attending to their duties as this is written: they have made no reference to the events. But it is things like that that make us Americans catch our breath and stand straighter!"

Property of
George de Tarnowsky
Colonel Med-Res, U.S.A.
30 N. Michigan Ave
Chicago, Ill.

The Germans Almost Win

Kaiser's Battle

February 1918

This ruthless devastation was not a figment of the imagination concocted out of whole cloth in order to create at home the proper degree of war hysteria; it actually took place, under official orders and I PERSONALLY SAW IT WITH MY OWN EYES!...Never did it enter this French surgeon's head to hasten death or omit any beneficial line of treatment; yet he was at any moment liable to lose liberty if not life, caring for bitter enemies. There is nobility of character in war-time!

The Germans almost win!

The American Indian of pioneer days relied on a multitude of signs for detection of approaching danger, but generally kept his eyes focused on the earth and moss and tree-bark around him. In the World War, an impending attack in force was widely advertised overhead and real surprise advances were few and far between. When a hitherto quiet sector, where existed, one might almost say, a gentleman's agreement to live and let live, suddenly began to receive heavy artillery shells manifestly directed towards bridges, railroad junctions, munitions dumps, and other military stores; when enemy planes started night bombing expeditions and dropped their highly unpleasant explosives rather indiscriminately over the landscape and our own anti-aircraft guns nightly kept up a furious but usually unsuccessful air barrage, officers and troops sensed a coming onslaught although neither official despatches nor new orders from H.Q. confirmed these suspicions. Late in 1917, French and British military leaders realised the possibility of a powerful German offensive in the near future. Thanks to the Russian revolution and capitulation of Roumania, many German divisions had been released from the Eastern fronts and sent to France. The enemy now had a clear superiority in numbers, and thousands of Russian and Roumanian guns were available to reinforce it's already formidable artillery. Although we declared war in April 1917 and our 1st Division landed in France late in June, the American army was not ready to enter the combat zone for many more months to come. The Germans shrewdly planned to separate the British from the French armies and defeat them piecemeal before our divisions could be of any assistance. Early in February the British agreed to extend their lines eastward and take over the sector hitherto defended by the French 1Xth army. While on one of my tours of investigation as the guest of General T....... in the tiny village of Ugny-le-Guay late in January 1918, British officers were already studying the defense system occupied by the 31st French army corps.

The British Vth army (General G....) quietly slipped in and occupied the rather weak system of trenches which their allies had judged sufficient to prevent a breaking through. When I visited the English C.C.S. established in the town of Ham in Picardy, on February 18th, the officers frankly admitted that their new lines were thinly held. They sensed impending danger and openly doubted their ability to withstand a determined assault. Although only a few members of the General Staff knew it at the time, the real object in extending the British lines was to enable the French to form reserve divisions which would be available wherever danger threatened. This of course the Germans knew through their superb spy system and, while they forestalled Allied plans to a certain extent, their attack came just a little too late to enable them to break through to a final decision. All of this precise information was denied us during hostilities, and it was only by comparing my war diary several years later with official reports issued by respective War Departments or Commanders-in-Chief that the true situation of March 21st 1918 could be unravelled. In wartime one obeys without asking why! Well do I remember the solemn interview which the same General T........gave us, behind closed doors, at the Ambulance des Alliés a few days after the French personnel had evacuated that hospital. "Messieurs les Américains", he said, "I need hardly ask you not to repeat a word of what I am about to tell you. Things are not going well with us just at present; l'heure est sombre - the hour is a dark one - and the final outcome is uncertain. We are rushing our reserves up as fast as possible and hope to stop the enemy advance created by the defeat of our British allies. We hope the situation will soon become more hopeful"

It was the last time I was to see the general. For various reasons one refrains from mentioning he had been recalled from his active command at the front and was henceforth, until further orders, to take charge of a group of camps for German prisoners. In the French army, the term "Limogé" was used to designate the fate of officers whose active services were, for one reason or another, no longer needed, Limoges being a very charming city in which to live in retirement and oblivion! In the American Expeditionary Forces, an officer who failed to come up to expectations or unsatisfactorily carried out a duty assigned him was sent home to the U.S. as "Instructor"!

As early as March 4th, 1918, my diary began to record unusual activity in our zone. Heavy shells began falling on Noyon, Ribécourt, and Compiègne; aerial bombs time and again just missed the railroad line following the Oise river or came crashing on the hospital grounds. During one particularly dark night, fourteen bombs fell in a straight line, just missing three of the wooden barracks which were used as wards for our wounded. Windows were shattered to pieces, floors danced up and down and, in the midst of the turmoil and frantic cries of the helpless, bed-ridden Poilus, Corporal Bonne, of whom I have written elsewhere, rushed from ward to ward, dark lantern in hand, bare-footed and long night-gown trailing to the ground and, in his soothing melodious voice - as though chanting one of his daily masses - beseeched the wounded to be quiet. "Calmez-vous, mes enfants, calmez vous: le Bon Dieu vous protège!" - calm yourselves my children, calm yourselves; the good God protects you! Dear good simple soul; how many times have I wished for his address in France! He was but one of thousands of ordained Catholic, Jewish, or Protestant rabbis or priests who, in the hour of peril, put patriotism above the cloth and served France in her hour of need. As soon as daylight enabled us to see, we inspected the shell-holes and found that all of them paralelled the railroad; it was evident that iron rails, and not red crosses, had been the air-raiders' objective. After all, why waste amunition on non-combatant Medicos or soldiers who, at least temporarily, are hors-de-combat? Throughout my wanderings along the front, I searched in vain for a bona-fide case of willful bombing of a hospital; whenever a shell or aerial torpedo had fallen on a first-aid station, Field or Evacuation hospital, some important military reason for the stray shell could be unearthed. Batteries of artillery would unlimber near our hospital grounds, fire all night, and depart for another location in the early morning; we received the counter-battery during the day! Munition dumps would be established in the vicinity of the hospital formation; no matter how well the piles of shells were camouflaged, their location would be ascertained sooner or later by the enemy. All branches of the service seek shelter for billets, depots, dumps, and P.C.s (posts of command). The plain fact of the matter is that, in war-time and notwithstanding the dictates of humanity, we, of the Medical Corps and the wounded under our care, are a nuisance to combatant troops. Our ambulances use roads sorely needed for the movement of troops, artillery, munitions, and supplies of all kinds; every medical establishment in the fighting zone is in somebody's way and even hospital trains carrying wounded to the base seriously hamper the rapid transportation of troops to the front or from sector to sector. It cannot be otherwise, and military surgeons had to learn to bow down to the inevitable, consoling themselves with the indubitable fact that, psychically as well as technically, they were an indispensible part of any army.

On March 6th, having had to go to the city of Soissons in order to get needed equipment for the wounded soldiers of our 26th Division who were ready for evacuation to the Paris zone, I was, confidentially of course, informed by the Division quartermaster that the New-England division was preparing to retire from the Soissons-Vailly sector. They were relieved by the French on March 18th, only three days before the beginning of the Kaiser's battle! March 8th a prolonged "dog-fight" occured over the hospital grounds, shrapnell bullets falling on roofs of barracks and all over the chateau park. My steel helmet and tip of right shoe were both hit. The bullets were presumably from French anti-aircraft guns, falling back to earth after failing to hit the invisible enemy overhead. Searchlights were not as perfect then as they became later on; the

German planes passed through the barrage and their bombs soon could be heard and seen falling on the city of Compiègne four and a half miles to the south of the hospital.

March 10th was my final tour investigation, a series of advanced first-aid posts along the famous Chemin-des-Dames overlooking the Aisne river valley. Little did I dream then that two and a half months later - on May 27th, to be precise - the Germans would crash through that sector and cross the Marne river at Chateau-Thierry before being halted!

♦♦♦♦♦

General Headquarters,
American Expeditionary Forces,
France.

4, March, 1918.

Special orders.
No 63

Extract

*　　*　　*

25. Major G.....T........, Medical Reserve Corps, will proceed to Craonne, France, to carry out the instructions of the Commander-in-Chief, and upon completion of this duty will return to his proper station.

The travel directed in necessary in the military service.

By command of General Pershing:

James G. Harbord
Brigadier General,
Chief of staff.

Official:
Benj. Alvord
Adjutant General.

♦♦♦♦♦

Armed with the above order and military identification card without which it was never healthy to travel from sector to sector during the war, I proceeded by train from Compiègne to Soissons and presented myself at H.Q. of Trains and Military Police of our 26th Division. The tortuous roads along the hillsides of the Chemin-des-Dames being rather difficult to negociate with an automobile,

Colonel S.....proposed that a side-car and driver be requisitioned for my trip. While waiting for my bumpy and jumpy conveyance to arrive, I had the good fortune to be able to excricate an adventurous but entirely innocent soldier from being arrested by the French secret service. An excited, middle-aged Frenchman whose military bearing belied his civilian disguise, ushered into H.Q. a thoroughly frightened boy of some twenty years of age. He had been picked up on a military train running towards Compiègne, thoroughly questioned in indifferent English by the French counter-spy and, not being able to give any other reason for his presence out-of-bounds other than he "wanted to see what the country was like", he was suspected of being a German spy! "I know this soldier is not telling me the truth!", said the Frenchman, "he says he is a student at the University of New-York: I only know of Columbia University in that city". "Colonel," said I, "there is a University of New-York; I know it's medical school and hospital". "Glad to know that, sir, glad to know that; let's ask the soldier to tell us about it". The boy did, most accurately; his "dog-tag" - as army serial numbers, stamped on metal and worn around the neck, are called - was displayed and checked up; a few leading questions regarding place and date of enlistment, ports of embarcation and debarcation, branch of service, and company he belonged to, and the French counter-spy was assured that "tout va bien" all's well. The innocent boy simply didn't know his Division boundaries and naively thought he might ride up and down the train in order to see the sights! Realising the possible gravity of his situation, the youngster promissed to be more careful in the future, clicked his heels, saluted, and left the C.O.'s office with a sheepish grin of relief on his open face. "Gad, sir" said Colonel S...."I'm mighty glad you were around; never heard of the U. of New-York before; didn't want one of our men to get in the hands of the French counter-spy system; our good friends are very nervous and jumpy these days; something's in the air, it's coming soon - God only knows where!"

The side-car trip was a blissfully peaceful one; hardly a shot was fired throughout the day, and only the distant rumbling of artillery disturbed the almost uncanny silence of that war-torn region. Looking at the churned-up land where every tree had been hit by shells and animal life, except rats, had totally disappeared, one idly wondered if vegetation would ever again spring up. Aid-posts and Auto-Chir's - Automobile Chirurgical or specialized surgical hospitals rapidly moved on trucks from sector to sector - were inspected at Chavannes, Craonelle, and Craonne, all occupying bomb-proof dug-outs carved out of the hillsides. The day's work ended at Beaurieux where the French Xth army had established it's advanced surgical center (Centre Chirurgical Avancé); all severely wounded cases were operated there before evacuation to the base. Beaurieux was later to be indelibly impressed on my mind for, on May 27th 1918 one of my best surgical nurses whom I had loaned to the surgical center, Miss C... W......., was captured by the Germans - but that is an epic in itself!

Returning to Soissons in the evening, I was informed that train service has ceased on account of night bombardments, and no American drivers were capable of taking me back to my hospital at night-time; we would get hopelessly lost or shot at by some nervous sentinel. "You will find a hotel of sorts near the main square" said Colonel S...., "the side-car will be awaiting you here in the morning; good night and...good luck". Off I went in the pitch darkness of the deserted streets; shells began falling here and there, and occasional footsteps could be heard clanging on the cobble-stones as officers or soldiers rushed from shelter to shelter, trying to reach their billets. Getting hopelessly lost, I stood under a door-way listening to the bombardment and crouching low at every flash. A perfect hit at a street angle illuminated the spot where I stood and, blessed Lady Luck, I salute thee!, right opposite me was the name of a hotel in gold letters! Memory for details fails me somewhat at times, but the name was something like the "Coq d'Or", the golden rooster. I dashed across the street and pounded on the door with hands and feet. After what seemed a long and anxious wait: "qui va la?", who goes there?, came a muffled voice from behind the enormously thick and nail-studded door. "Officier Américain qui cherche un abri!", American officer seeking shelter. Chains and bars were let down, a huge lock was turned, and the old hotel owner beckoned me in hastily, slammed, and bolted the door and, pointing the narrow and curved stairway on one side of the hallway: "Aux caves, Monsieur, vite, vite!" to the cellars sir, quickly! Gingerly picking my steps down several flights of a circular and rather narrow stairway, I found myself in the dimly lighted wine cellars of the hotel with a heterogenous group of men and women, who had been ordered to remain below until the bombardment and air-raids had ceased. An American Red-Cross nurse sat on top of a huge wine cask, swinging her feet and chatting with a junior Infantry officer; two "Y" men paced up and down nervously, evidently new to battle life; several French soldiers and a Major of Artillery, the latter smoking incessantly and chewing at his grey mustache, Madame la Patronne (proprietress) with her husband and civilian help completed the marooned crowd. Smoking, drinking, chatting, and yawning - principally the latter - we remained underground until four in the morning when, after twenty minutes of silence overhead, the landlord allowed us to go upstairs and occupy our beds. Bed-bugs and "cooties" made it an all-night affair for me, what the French ironically call a "nuit blanche" or white night! It was a relief to be able to get a cup of coffee, sweetened with saccharine, and a piece of French war-bread, the latter apparently made out of everything except flour, and report at Military Police H.Q. The cool ride down the Aisne, across Compiègne, and up the Oise valley to my hospital was delightful. Hurriedly typewriting my report in triplicate, one copy for the Chief Surgeon's office, one for the publisher, and the last one for myself, I caught the evening train for Paris and mailed my final chapter. It had been a case of hurry, hurry, hurry, from the moment I undertook the assignment in Washington in July

1917! Some of my stay-at-home friends sneeringly refered to my experiences as "joy-riding" in Europe: I wonder how they would have stood up under the strain?

A week's leave-of-absence, spent in Nice with a beloved sister was a heaven-sent rest and relaxation for me, but my arrival in Paris on March 21st coincided with the beginning of the great German offensive. Tremendous concentrations of artillery fire could be heard all day long. The afternoon train for Compiègne did not leave the Gare du Nord before eight o'clock in the evening; flashes of light and heavy explosions to the north of us made everybody realise that a real battle was on, although the afternoon official "communiqué" had only curtly mentioned "increased artillery activity in the Cambrai-St Quentin sector". An air-raid over the town of Pierrefittes held the train up for over an hour; Creil, a very important railway junction, was a tangled mass of confusion, trains trying to pull out, others trying to pull in, car after car of French soldiers sidetracked here, there and everywhere; individual Poilus jumping out of cars and dashing towards the station cantine, to be sternly ordered back to their units by M.P.s; smouldering fires and clouds of smoke from previous explosions, "hell let loose with a vengeance"! Miles before we arrived at Compiègne, we could see the city, brilliantly illuminated by a monstrous sheet of flames, which shot upwards like a volcanic eruption. A German aerial bomb had registered a direct hit on an important gazoline depot, and thousands of five-gallon cans had exploded. The station cantine had been hit; it's proprietor and several French soldiers killed. The tracks around the station had been repeatedly hit, and frantic attempts were being made to detour all military trains going north. No hospital car in sight, heavy musette and duffle-bag to carry four and a half miles! Fortunately, I had spied General T.....and his wife on the same train; they most graciously made room for me in his staff car. Had it not been for my luggage, it would have been much quicker to have walked to the hospital. The main road to Noyon was a mass of French troops of all branches, convoys of all sorts and descriptions, batteries of 75's and munition trains tramping northward. Not daring to question the general, I leaned out of the car window, fascinated by the sight and sound of devoted patriots hurrying up to stop..what? Leaving the main road at the village of Longueil-Annel, we soon reached the Ambulance des Alliés; it was then four in the morning. What to do with the wounded American and French officers and soldiers in the hospital was our first concern. How they were safely evacuated and how the hospital continued to function until May 27th, 1918 will be detailed elsewhere; the battle's the thing now!

With the advent of British wounded, we were able, piecemeal, to get a picture of the terrible onslaught called the Kaiser's battle. At four in the morning of March 21st, 1918, the Germans began the drive by opening from thousands of guns of all calibers a bombardment surpassing anything that had gone before. Poison-gas shells were used in profusion while, at the same time, long-range guns swept the roads, railway junctions, and possible concentration points in the rear, preventing the British from sending up reinforcements. It was reported at the time that sixty-four German divisions crashed against eleven British; the odds were truly overwhelming! A heavy mist helped the invaders and, after several hours of artillery preparation, the Feldgrau soldiers of the Kaiser were almost upon the British front lines before they could be seen. Resistance, though heroic, was futile; the British had to give way. Some days later, a French officer whose life we managed to save, in spite of the ghastly abdominal wound which he had received, confided to us that - so nearly had the enemy succeeded in separating the French from the English armies - for a few hours there actually was a gap fifteen miles wide without any Allied defenders in it! The invaders apparently suspected that a trap had been laid for them and hesitated in their advance. A rapidly improvised unit which included American engineers, labor bataillons, and British stragglers who had lost contact with their units, managed to plug up and successfully defend the gap until reinforcements came up. What a narrow squeak!

There is no sadder sight than that of tangled-up remnants of a beaten army retreating without any possibility of military cohesion or organization. Singly or in groups, British infantrymen, artillerymen, engineers, and batmen passed down the road in front of our operating rooms or stopped for a bite to eat and a little well-earned rest. Officers knew they had been overwhelmed and were correspondingly disheartened; the soldiers, with rather characteristic cockiness, would assure us that "we are not retreating, sir; the French are releaving us in our old sector, and we have orders to reassemble in the town of Estrées-St-Denis." Quite true, as far as name of town and French relief were concerned! Even after reforming what was left of the unfortunate British Vth Army, it was not destined to have much better luck two months later; some of their elements were again swept away by the Chemin-des-Dames surprise attack.

According to my diary, the last British straggler to enter our hospital arrived on March 29th, eight days after the beginning of the Kaiser's battle. Still suffering from trench fever, the poor soldier only had a hazy recollection of the battle and miraculous escape from capture. Vaguely he remembered having a high fever and of being sent to a C.C.S. in the city of Péronne, of being taken by ambulance or motor truck from one temporary shelter to another, and finally being classified as a "walking case" and told to march south towards Estrées-St-Denis. Fellow-stragglers had shared their meager rations with him and passed on forward, leaving him to follow down the eucalyptus-lined Military French roads. With true animal instinct, whenever the chills and fever came upon him, he would crawl under some shelter and hide until his head cleared once more, and he could stagger on in the right direction. A few days' rest in the hospital, and we were able to send him on to his own unit. Henceforth, all our wounded were French with a few Germans. The latter would not at first believe that American officers could be in France for, they said, they had been told that Admiral von-Tirpitz's submarine "schreklichkeit" had absolutely prevented any Americans from crossing the Atlantic. This "official deception" was to have a very depressing effect on the morale of the German troops a few weeks later when Cantigny, Belleau Woods, and Chateau-Thierry informed the world that the A.E.F. was in France! From the type of wounded we received, one of two deductions could be made: either the best shock troops were being held back for some later battle, or German man-power was fast reaching the point of exhaustion. Physically, omitting of course their recent wounds, few of them would have been accepted in our training camps at home; they were under-nourrished, undersized, and their ages varied between thirty and forty-five years of age. All of them belonged to reserves of the second or third class. It saddened one to notice how suspicious these poor wounded enemies were of our surgical ministrations. Just before being anesthetized, a highly intelligent Sargeant of Infantry on whose right arm I was going to operate, asked for permission to speak to me. "Bitte, Herr Major" said he, gazing intently at me with his piercing steel-blue eyes, "are you going to allow me to awaken from the Ether?" Not immediately grasping the true significance of his question, I assured him that the operation, though a serious one, was not a particularly dangerous one, merely the removal of pieces of shell fragments and setting of a fractured bone, and that his chances of recovery and preservation of arm were good. "Ja, Ja, Herr Major, I think I can recover from my wounds but, are you going to let me live?" Murder by anesthesia! The thought was preposterous, horrible, nauseating! All the Anglo-Saxon traditions of true sportsmanship under which I had received my early training arose in protest against such a base suspicion. With all the earnestness that the circumstances and surroundings demanded, I gave the poor misinformed Feldwebel my word of honor that the anesthetic and operation would be carried out with exactly the same care and technique which we used on our own men! The operation was complete success and, as the convalescent German left our hospital for a war-prisoner's camp infirmary, he thanked me mutely with his eyes. Had he been in a military hospital of the A.E.F., we might have become better acquainted but, in a French formation, the situation was a rather delicate one. The Feldwebel's experience must nevertheless have convinced him that Americans were not wild Apaches! While there undoubtedly existed a sullen hatred of the Boche in every Frenchman's heart, largely caused by the wanton destruction of all vegetation, including fruit trees in the old Hindenburg salient of Picardy, the treatment of their war prisoners in our hospital was correct if not kind. One always feels like reminding Americans who express surprise over the French post-war dislike and suspicion of Germany and German motives that our country was not invaded, our coal mines flooded, towns destroyed, and countryside razed to the ground in order to form what the enemy complacently called a "glacis" for future assaults! One fancies that, if our farm wells had been deliberately polluted, farm houses razed to the ground, hot-houses painstakingly smashed to the last pane of glass, fruit-trees, vines, and berry-bushes systematically chopped down a few inches from the ground we, also, MIGHT REMEMBER! This ruthless devastation was not a figment of the imagination concocted out of whole cloth in order to create at home the proper degree of war hysteria; it actually took place, under official orders, and I PERSONALLY SAW IT WITH MY OWN EYES! When the French Evacuation Hospital of Royallieu, near Compiègne, was ordered evacuated, a number of severely wounded Germans were left behind in charge of a junior medical officer and necessary orderlies. This young Frenchman's orders were to remain at his post as long as any of the wounded lived and to be taken prisoner, should the enemy succeed in crashing down the Oise valley from Noyon and Mount Renaud. If, on the other hand, all of the prisoners were to die, the surgeon and his men were to set fire to the hospital and retreat on the city of Creil. When I visited him, he showed the strain under which he had been carrying on but, pointing to the main road leading to Compiègne, kept repeating the famous Verdun challenge: "Ils ne passeront pas!" - they shall not pass! He asked me to examine his patients; all were doing well, all were out of danger when later evacuated to the rear. Never did it enter this French surgeon's head to hasten death or omit any beneficial line of treatment; yet he was at any moment liable to lose liberty, if not life, caring for bitter enemies. There is nobility of character in war-time!

In the later phases of the Kaiser's battle, aviation played a more and more important role. So eager were the French airmen to do their part towards preventing the Germans from coming down the Oise valley from Noyon and Mount Renaud, that they began flying lower and lower over enemy trenches, using their synchronized machine guns with deadly effect, but also paying dearly with their lives. Many of them, though severely wounded, managed to keep their planes going until they were safe behind our own lines. One of these planes fell in a field immediately in front of the hospital, turned turtle and, through some quite unexplainable miracle - the ignition switch not having been shut off - failed to catch on fire. One of the orderlies saw the fall and gave the alarm. We rushed out and extricated pilot and observer. The latter only had a machine-gun bullet through his right cheek, but the pilot's chest had been pierced through and through. The plane had been sent out on a tour of artillery regulation or "réglage", signalling the batteries just where their shells were falling and correcting faulty direction. The pilot, shot through the right lung and bleeding internally, managed to turn his plane towards the south. The observer, sitting behind him suddenly noticed that the pilot had fallen forward in a dead faint and, glancing ahead, realised that the plane was heading for the earth with a line of tall trees a few hundred yards ahead. With rare presence of mind, the boy - he was barely twenty-one years old - leaned forward and pulled the controls towards him. In the nick of time, the plane rose slightly, missed the tops of the trees by inches, and crashed down to bury it's nose in the soft earth. Just what killed the engine at that moment was hard to explain. Among my treasured war souvenirs is a "swagger-stick" fashioned out of a piece of the propeller, ornamented at both ends with hand-beaten silver, which one of the wounded soldiers presented to me as a token of gratitude. The pilot went through a stormy convalescence but ultimately recovered; the observer's wound healed in a few days. We witnessed another thrilling aeroplane epic a few days later. The captain of a neighboring escadrille was killed by German anti-aircraft bullets while on a tour of observation. As the plane began it's deadly spin to earth, the observer seized the controls and made a beautiful landing in that same field of ours. Henceforth, to us, it became known as the Ambulance des Alliés landing field! No less than six French planes made unforseen use of it in life-saving descents.

Each time an aviator was buried in the hospital cemetery, Captain Louis Meiffre,C.O. of air-forces in the Vth Corps area, would attend the funeral. Many staff officers would also turn up and, overhead, planes would circle around, dip down gracefully in last farewell, and throw flowers in the general direction of the grave. At one of these funerals, German aviators had the bad taste to appear in the distance. In an instant, every French plane afloat started in pursuit; the enemy declined to stage a "dog-fight" and retired behind his lines. Captain Meiffre attended six aviator funerals: on May 2nd, 1918, we all attended his! He had become endeared to all of us by the simplicity of his nature, devotion to the members of his escadrilles, calm indifference to danger, and atmosphere of optimism which he always created around him. Speaking excellent English, extremely handsome, and every inch a man of the world, the nurses adored him. Without saying a word to anybody, they scoured the Gannelong hill and gullies for mariposa lillies, anemonies, and early violets, gathered an armful of ferns, and fashioned the lot into a beautiful cross which they laid on the casket with a card on which they had written "From the American officers and nurses to a brave gentleman". Four French generals with their staffs gathered around the grave while dear Corporal Bonne, his priestly vestments thrown over his war-worn reservist's uniform, officiated. Standing there, cap in hand, the thought suddenly struck me: "what an opportunity the Germans have to make a direct hit and destroy some of the best military brains of France!" Fortunately, no enemy planes were sighted on that day. As the morale of the Vth army corps, escadrilles was considerably lowered by the loss of their C.O., they were removed to a quiet sector. Before departing, and as a delicate token of their appreciation for our surgical services, one of their planes, carrying a photographer, circled over the hospital and took a panoramic view of it. The two huge red crosses laid on the ground near the wards showed up very plainly, but, alas, the main road and railway line paralelled the lower end of the chateau gardens and constituted a very legitimate military target!

Only once did we see a German plane which was not a mass of tangled wires and wings. A D5 type Albatross whose pilot had been killed by two young French aviators nose-dived directly into one of our second-line trenches, where it remained an object of curiosity until removed to a war museum. Aside from the propeller blades which were smashed and a slight tear in the wings, it was hardly damaged. A photograph of the trench and plane was given me with the caption: "Albatros D5, Descendu le 23/4/1918 par l'aspirant Barbaud et le caporal Guillaume".

Of the actual battle, or rather series of battles, retreats, and advances, what can a Medical Officer describe as having been an actual witness to? Little or nothing! True, one could witness artillery duels, see shells exploding, and listen to the roar of guns until one's ear-drums ached; shells did fall around the hospital, and bombs narrowly missed our Chateau and wooden barracks. For actual details, we had to question the wounded and they, in turn, could only describe what had occured a few yards to the right and left of each individual. Battles in their ensemble can only be witnessed at a distance, field glasses in hand, and from some elevated post-of-observation; scrutiny of the topographic map laid out before us, and knowledge of axes of rotation of the two opposing forces can alone enable one to foretell the probable result of an engagement. That General Pellé could find dead Germans piled up seven high on the Lassigny plateau or on the Plémont, meant that these were two key positions which the enemy had to conquer and consolidate before he could safely advance; that Mount Renaud changed hands almost daily meant that the Oise valley could not be used by the Germans in their anticipated march through Compiègne, Créil, and Chantilly before reaching Paris, until the French had definitely been driven from it. When, by the first week in May 1918, all key positions were firmly in French hands, we knew that, almost winning, the Germans had lost the battle. When, from May 10th on, long-range shelling and activated air-raids manifested themselves in the Compiègne-Soissons sector, we sensed another attack, in a new direction. Where would it fall this time? "On va bientôt se chamailler ailleurs.", there's going to be fighting elsewhere, soon, said our wounded. Where? A silent but eloquent shrug of the shoulders, such as only a Frenchman can give, was their only answer. As heavy shelling and bombing of the rear zones increased, the number of wounded received at the hospital decreased. The front, in our sector, was again stabilized. That the higher command sensed an impending onslaught elsewhere was indicated by the removal to Pont-Ste-Maxence of the French personnel of our hospital and the appointment of Major J.J.M. as C.O. of a new American Red Cross Military hospital at Beauvais. French hospitals at Villers-s-Coudun, Cugny, Compiègne, and Royallieu closed up and departed. By May 20th, we knew that the Ambulance des Alliés would no longer functionate after the 27th, and, on that day, the few remaining nurses and I took the train at Compiègne and reported in Paris for a new duty. Preliminary surveys of the Auteuil race-track in the Bois de Boulogne had been previously made, and another Red Cross Military Hospital, to be known as No 5., was rapidly organized under my command.

Why have I dwelt on a phase of the World War in which our Expeditionary Forces took no part? Primarily to remind my friends at home of the bravery, fortitude, and tenacity of our French comrades; secondarily to correct the utterly false impression that "we won the war!"! There is and was enough glory to go around; Belgians, British, French, and Italians also played an important part in the final phases of the stupendous conflict. Russia, up to the time of her revolution, kept hundreds of Austrian, Bulgarian, Turkish, and German divisions diverted from the French front. We came in late, filled gaps at crucial points, fought gloriously, and won many battles, BUT WE DIDN'T WIN THE WAR!

Captured! The Epic of a World War Nurse

Sunday, May 26, 1918

The capture of an unwounded War Nurse however, was so rare, so unexpected as to warranted being reported.

Prisoners of war, wounded or unwounded, were insignificant details in the gigantic conflict. Wounded men could not prevent being captured whenever, in the course of a local surprise attack, a rapid advance or retreat, there was no opportunity for the removal of battle casualties. Of the unwounded prisoners, a few were deserters: the vast majority had surrendered when, in the opinion of the senior commanding officer or warrant officer, resistance in that particular sector had become futile. The capture of an unwounded War Nurse however, was so rare, so unexpected as to warrant being reported.

Mention has been made elsewhere of the surprise attack which, beginning north of the Chemin-des-Dames sector shortly after midnight of May 26th, 1918, enabled the Germans to cross the Aisne and Vesle rivers, sweep south to Chateau-Thierry, and cross the Marne to Vaux village and Belleau woods before being halted by Franco-American troops. At the time of the attack, this sector was rather thinly held by British and French divisions, which had been sent there to rest up after their arduous campaign in Picardy. It may be unfair to hint that the morale of these tired troops was not of the best, but facts are facts nevertheless; units which have recently been overwhelmed by sheer force of numbers and concentration of artillery fire and obliged to retreat, oftentimes in disorder, are not capable of becoming shock troops once more without a thorough shaking up of staff officers.

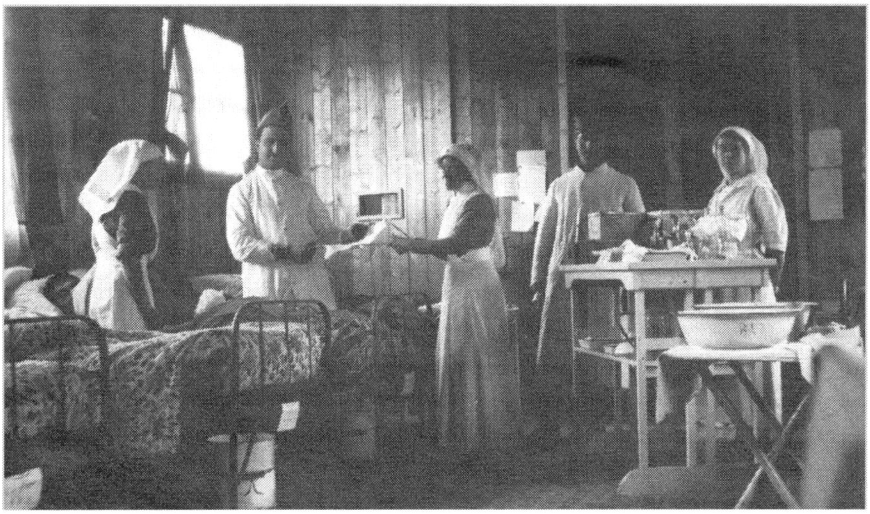

C...W.......a few days before being captured. She is seen standing beside the surgical table.

In conversation with a group of French line officers shortly after the Armistice, it was claimed by several of them that the German advance on Chateau-Thierry had not been a surprise attack, but that, on the contrary, Marshall Foch had deliberately weakened his lines along the Chemin-des-Dames in order to create another German salient, which might later on, like St Michel, be attacked simultaneously on either side and from it's apex. "Improbable, but such an opinion can be defended," replied one of the less chauvinistic officers of the group. Knowing as I did that the French had lost some forty thousand hospital beds with their buildings and equipment in the Chateau-Thierry salient and that our own American army hospitalization program had been badly dislocated thereby, I could have advanced many telling arguments against the hypothesis of Marshall Foch's supposed strategy on this occasion; wisdom and proper regard for sensitive Allied feelings kept me silent. Several years later, it was interresting to discover, both in General Ludendorff's and Marshall Foch's memoirs that the Chemin-des-Dames attack had been indeed a complete surprise. With justifiable pride, Lundendorff boasts that: "Concentration of forty to fifty divisions had not been observed by the enemy, nor had it been reported to him by his highly developed secret service. The marches took place at night, but the troops sang as they passed through the villages. Enemy airmen did not discover the railway transportation that had been taking place behind the front of attack. The guns and trench mortars with their ammunition were in position behind the foremost trenches and, in places, even in front of these trenches. The divisions were crowded together in anti-aircraft shelters behind the jumping-off places in the front line." Foch, sensing another impending peril, yet unable to discover from whence it would come nor when, thus describes his perplexities:

"To the battle period, the opening of which had been marked by the German guns on March 25th, 1918, there had succeeded a period of calm. We knew that the enemy had from seventy-six to eighty divisions at his disposal. Where would they suddenly announce themselves? WE SEARCHED THE HORIZON BUT THE MYSTERY REMAINED UNSOLVED! Such was the situation, when suddenly it was learned that a new German offensive was on the point of being launched. On the afternoon of May 26th,1918, German prisoners declared that an attack on the Chemin-des-Dames would take place that night after a short artillery preparation, which would begin at one A.M. This information was correct."

The advanced surgical center of Beaurieux in the Chemin-des-Dames sector received both British and French wounded; it was quite natural therefore that one of the best surgical nurses of the Ambulance des Alliés should have been "loaned" to this group of hospitals, especially as she was a good bi-linguist. Miss Colt W......., a British nurse who had seen considerable service in France, volunteered for the post and took up her new duties about May 15th.

The Oise Canal
Where Frenchmen fish and fish, but never get a nibble!

A graduate of one of the best English hospitals, C.W. was one of many hundred similar adventurous spirits who had heeded the call to duty, been sent to France shortly after onset of hostilities, and assigned by the Red Cross to various French military Base Hospitals. Ultimately, she had gravitated towards the front lines and been assigned to the Ambulance des Alliés hospital as surgical nurse. Capable, reticent, flitting from Ward to Mess Hall to Nurses quarters, occasionally taking a short stroll in the vicinity of the hospital grounds with one of the Junior French surgeons, or accompanying him on one of those classical fishing expeditions along the Oise Canal where Frenchmen fish and fish and fish....without ever getting a nibble or bite, Colt W.......remained a stranger to the American surgeons. Blonde, with the British complexion, which is the unspoken envy of American girls, she was, without being a beauty, quite comfortable to look at. The wounded "Poilus" affectionately referred to her as "la blonde Albion" or "Miss Anglaise". Who her people were, whence she came from, what her social status at home was, nobody apparently knew. The reticence often observed during the war reminded one of frontier days in our West, where asking a man what his name was and where he came from often resulted in a shooting fracas! While she was able to understand French sufficiently well for nursing or operating room duties, it was self-evident that no extensive quizzing on the part of the favored French surgeon could possibly be carried out, the latter's English vocabulary consisting in one single word, "Miss"! Without ever attempting to break down the unmistakeable barrier which Colt W.......had all too manifestly elected to erect between herself and the English-speaking members of the hospital, we had come to the conclusion that she did not wish us to know anything concerning her past life and experiences which, we felt certain, included a romance so precious, unusual - possibly unconventional - as to preclude outsiders sharing in it's knowledge!

So quiet was the Chemin-des-Dames sector in May 1918 that it was almost looked upon as a rest area for tired troops! The thirty-three mile front was manned by only ten French and British divisions, four French and three British divisions in the front lines with one British and two French divisions in reserve. How the Germans managed to bring up nearly half a million men, enormous concentrations of artillery with necessary shells, and special supplies and equipment for the shock troops without arousing Allied commanders' suspicions, has remained a mystery to this day. It was currently rumored at the time - one of those army rumors commonly referred to by a Latin word, which is no longer used in polite circles - that a French general had committed suicide rather than face a court-martial; if true, the details of this tragedy have been carefully hidden in the secret archives of the French War Ministry! From German prisoners, we heard that the preparatory advance had been carried out by a careful process of infiltration. No troops or munition trains had been allowed to move by day-time, horses' hoofs were silenced by wrappings of straw and burlap under their shoes, men, trucks, horses, and artillery were hidden in woods or camouflaged so that aviators failed to report any unusual concentration of troops behind the threatened sector.

During the early morning hours of May 27th, Miss W.......was awakened once or twice by the booming of artillery fire but, seasoned war nurse that she was, she covered her head with her pillow and resumed her slumbers, convinced that nothing unusual was happening. Reveille, breakfast and morning ward, and operating room duties went on normally; the heavy firing had ceased during the night. At nine o'clock in the morning, an excited French hospital orderly rushed into the ward where Miss W.......was doing surgical dressings, shouting: "Les Boches sont ici! Les Boches sont ici! the Germans are here!"! Hardly had his warning cry ended when in strode a German Colonel, Medical Corps, followed by several junior surgeons! This particular ward or "Barraque Adrian" had been added to the older group of wards and happened to be in front of the Administration office, hence C.W...had the doubtful honor of being the first nurse of the Beaurieux hospital center to be interviewed by the victorious enemy! "How terribly thrilling!" exclaimed one of the nurses who had met C.W......at the Gare du Nord in Paris when, the war over, she had been freed from prison camp and sent back to France, where she reported at British Red-Cross H.Q., "weren't you jolly well frightened out of your wits; didn't you scream, faint, or do something unusual?" "It happened all too suddenly; I didn't have time to think; fear did not seize me then - as it did later on, many a time; fear of poison gas, of hunger, of surgical infection, working as I had to without proper means of asepsis or antisepsis, fear of the officers and men around me, women hungry; later on, yes, I knew what fear was - but not on May 27th!" Thus did the poor girl answer some of our questions in a voice weakened by insufficient food and subsequent illness. "When the German Colonel spoke to me in French, brusquely asking me who was in command of the hospital I - automatically it seems to me now - turned to the scared orderly saying: Pierre, vas chercher Monsieur le Médecin-Major; dis-lui de venir ici tout-de-suite." Excitement was in the air by that time; nobody was allowed to leave the Centre Chirurgical. The German surgeons visited the entire hospital, jotted down the total number of vacant beds available for their own wounded, picked out the English and French wounded who, they decided, were capable of being transported to German hospitals in the interior, grouped the intransportable ones in certain wards so that Germans would not have any enemy wounded under the same roof and informed C.W.......and her co-nurses that, for the present they would remain on duty at the hospital. Surgeons, orderlies, and help were told to pack a few of their belongings and be ready for internment in a military camp. From May 27th to July 18th, the prisoner-nurses continued with their routine duties, harshly treated, poorly fed, and lodged, but otherwise not molested or insulted in any manner. Heavy and prolonged artillery fire from the general direction of Soissons, during the early morning hours of July 18th, and increasing uneasiness on the part of the German personnel, led Miss W......to surmise that the Allies were turning the tables on their enemies. (As a matter of fact, the Allied counter-offensive of July 18th marked the turning point of the war: after the abortive German offensive against the Rheims salient, the Allies maintained the offensive until Armistice day!). On July 19th, the German C.O. ordered the nurses to pack up and be ready to leave in an hour's time. Instead of being taken into Germany and interned in a war-prisoner's camp in accordance with the International Red Cross conventions, these nurses were merely shifted from one Field Hospital or main dressing station to another as the Germans slowly and stubbornly gave way, fighting rear-guard battles and salvaging what they could from the immense supplies they had poured into the Marne salient in confident preparation for a final onslaught on Paris. To keep nurses in the battle zone after having been made prisoners is almost unbelievable, yet, of the truth of this particular incident, there cannot be any doubt. When C...W.......recounted her experiences, it was with the calm detachment of one who already had a presentiment that the call to the great beyond was not far away. Of hatred for the enemy, she exhibited not a trace; listening to her story, observing her pale, anaemic, expressionless face, one received the impression that the treatment she had been subjected to had forever deprived the poor girl of the capacity for any further joy in life. Gone was her "joie de vivre", gone her former belief in chivalry, in the protective instinct of man which every normal-minded woman counts upon in life's daily contacts. Nothing remained, nothing mattered and, while at last free to come and go as she pleased, all illusions had been shattered, faith in humanity crushed, and, as far as she was concerned, life had been weighed in the balance and found to be not worth-while bothering one's head about!

From July 19th to August 8th, C...W.......carried out her enforced surgical duties in the furiously contested neck of the Marne - or Chateau-Thierry - salient. We Allies were fairly deluging this area at all hours of the day and night with high-explosive shells of all calibers, dropping gas-shells by the thousand, bombing by means of aeroplanes at night and Stokes-Mortars during the day, capturing, losing, and retaking such villages as Fismes and Fismettes, Craonne and Craonelle, before compelling the enemy to retrace his steps beyond the Chemin-des-Dames, down whose slopes he had marched so confidently on May 27th.

It is quite probable that the removal of German military stores so blocked transportation that their Dressing Stations and Field Hospitals could not follow the retreat fast enough. Day by day the nurses found themselves working nearer and nearer to skirmishing lines, always under bombardment, sleeping - when allowed to - in wretched dug-outs or cellars of destroyed buildings, poorly and irregularly fed, with no opportunities for laundering their meager change of underwear and uniform. "Nobody bothered about providing us with any sort of 'conveniences' such as any decent Anglo-Saxon woman is entitled to; we felt like wild beasts yet preferred to...injure our health rather than give up what little sense of modesty still remained in us!" Thus C...W.....essayed to depict what, to her, represented the lowest depth of the wretched existence she led while on duty as a prisoner nurse in the German army! It was quite evident that medical supplies could not be replaced as fast as requisitioned. The nurses used rusty old instruments, which had neither been sterilized by boiling nor kept in antiseptic solutions; dressings were scanty, splints improvised with any material at hand and often tied to broken limbs with twine, because bandages and adhesive tape were no longer available. "I wish I could find out how many of the wounded I attended to had their wounds become infected after receiving first-aid!" exclaimed C...W......musingly, "all my hospital training, all my knowledge of asepsis and antisepsis went for naught - we were savages, savages!"

Belonging as I do to profession which, in spite of certain modern tendencies towards commercialism and quantity, rather than quality, services to the public, is still and will continue to remain a noble and honored one, I hesitate somewhat to relate the conduct of several enemy Medical Officers on duty with C...W........and her small group of nurses in the village of Fismes. With artillery shells falling around them everywhere and desperate hand-to-hand encounters being fought a few miles south of Fismes, the number of wounded soon exceeded the capacity of the underground Field Hospital, which French engineers had excavated and the Germans were using. While the cowardly surgeons treated their wounded in this absolutely bomb-proof shelter, they ordered the poor nurses to go outdoors with their instruments, antiseptic solutions, and dressings and attend to the overflow of wounded lying on the ground. Many of these unfortunates had been laid down near the railway tracks which, of course, were receiving the severest barrage our Doughboys could lay down. For three days and nights did these heroic young women brave death in order to give help and comfort to wounded friend and foe alike. They probably snatched a few hours' sleep and gulped down some food at irregular intervals; by some providential miracle, none of them were wounded, but the exposure, loss of sleep, and fatigue initiated a disease which, ultimately, proved fatal in the case of C...W.......! Had the nurse not told me personally of her experiences, I would have refrained from jotting them down in my War Diary. That this incident was not typical of the conduct of German Medical officers during the war, goes without saying but, should any of these renegade members of my profession ever happen to read these lines and recognize themselves, I shall feel quite happy! Shunted from Fismes to Mont St Giles and, later, to Montigny-s-Vesle (which, in plain English, means the village of Montigny on the banks of the Vesle river), the same incidents were repeated: there could be no doubt about the cowardice of the three or four Surgeons these nurses had to work under. Having, in the early months of 1918, visited the Chemin-des-Dames sector a number of times, thanks to the kindness of my old friend, the late General Pellé, I was able to follow C...W.......'s descriptions, recognize the various hospitals she had passed through, and visualize the dangers she and her companions had been brutally and unnecessarily subjected to.

Had Cupid not bestirred himself at the right moment, one of my own sisters would, in all likelihood, have been captured during that same surprise attack of

May 27th! Visiting France in the summer of 1914, she had been caught in the maelstrom of French mobilization and, filled with energy, fired by the contagious enthousiasm, which permeated the entire French nation, had voluntarily joined the vast group of French women and foreign sympathisers called the "Union des Femmes de France". In due time, many of these devoted women became expert war nurses and were sent to various surgical centers back of the main system of defenses, which characterized the first three years of the World War. While she was on duty as surgical nurse in the H.O.E. (Hôpital Origine Étapes, or Evacuation hospital) at Montigny-s-Vesle, I had visited her on several occasions and noted a brilliant French surgeon who seemed particularly anxious to make me like him - an easy task for we became good friends after our first meeting! The romance progressed rapidly, as war romances usually did! Early in May 1918, my sister announced her engagement; her fiancé at once insisted that she accept duty in a less dangerously exposed sector - which she did! He remained, was transferred to Mont St Giles just as the German avalanche descended on the sector, operated on wounded Frenchmen until five in the morning and then, under orders, marched south with the rest of the staff - minus the unlucky junior surgeons selected to remain behind and be captured with their wounded - until he reached Chateau-Thierry and crossed the Marne before the enemy could catch up with him! How did the romance end? Oh! yes, of course! Why, they were married on July 4th, 1918, were allowed a twenty-four hour honeymoon and lived happily ever afterwards! My sister and brother-in-law were both able to corroborate the descriptions of hospitals and landscape contained in C...W.......'s story: it rang true and is true.

When, after the "Black Day" of August 8th and the still blacker days following August 20th, 1918, the German supreme command realised that they had lost all hope of winning the war, the prisoner-nurses were finally taken to Germany and interned in Camp Ruhleben. Of the railroad journey into Germany, which lasted several days, three facts had been indelibly impressed on C's mind: The nurses were kept locked up in a freight car minus bedding, light, or "conveniences"; food and water were handed to them through the partly opened sliding door of the car, just as though they had been wild beasts; whenever there was a delay at a junction point or regulating station, the car door would be pulled wide open, and guards would order the girls to stand in front of it so that German civilians could gaze and jeer at them. Fortunately, none of the nurses could understand the language - at the time!

Of her life in prison camp, C...W.......said little beyond stating that she was in the infirmary a good part of the time, suffering from a severe and intractable intestinal disturbance associated with hemorrhages in which sapped her strength. Prison regulations were harsh but not deliberately unkind; the food was consistently inadequate both as to quality and quantity and special diets, such as C.W. needed, were unavailable.

As the German officials at home gradually sensed the impending defeat of their country, so did their attitude towards prisoners of war change - for the better. The quality and quantity of the daily food ration improved, more freedom was allowed, the prisoners and flowers were even brought to the girls by the civilian population. "They all tried to make us forget what we had been through," said C.W. with a wan smile, "but it was too sudden and too thickly put on to fool us!"

A few days after the Armistice, the nurses were placed in the only first-class compartment of a repatriation train, supplied with ample food for their journey into France, and literally showered with bouquets of flowers which German girls threw at them through the open car-window as the ex-prisoners murmured a fervent "good-bye" in response to the shouts of "Glückliche reise, Schwestern; höffendlich, auf-wiedersehen!"

France did what she could to show appreciation and gratitude. The courtyard of the Invalides in Paris has been the scene of countless ceremonies - called "Prises d'Armes" - such as the one I witnessed late in November 1918 when C...W........received the coveted Croix de Guerre. She stood in line with a brave soldier or two, several officers, and a number of civilians, showing no emotion, expressing neither pleasure nor displeasure, indifferent, unfeeling, hopeless! "A charming gesture wasted," thought I, standing at attention in one of the galleries overlooking the famous "Cour d'honneur". A few weeks later, the poor girl was dead, among her own people in England; another nursing heroine!

A typical "Prise d'Armes" in the Court of Honor of the Invalides, Paris.
(The nurse in the picture is not C...W......but the author's sister!)

Ministry of Public Instruction and Fine Arts

Officer of Academy

November 20, 1919

E-0793.

FRENCH REPUBLIC

MINISTRY OF PUBLIC INSTRUCTION AND FINE ARTS

The Minister of Public Instruction and Fine Arts,

In conformity with Article 32 of the Organic Decree of March 17, 1808;

In conformity with Royal Orders of November 13, 1844, September 9, 1845, and November 1, 1846;

In conformity with Decree of December 9, 1850, April 7 and December 27, 1866, December 24, 1885, and August 4, 1898,

Decrees:

Lieutenant Colonal George A. Tanowski, Medical Corps, American Army, is named Officer of Academy.

Done at Paris, November 20, 1919.
Minister of Public Administration
and Fine Arts.
Signed: L. Lafferre.

True Copy:
Chief of Cabinet,
Albert Boucher.

Letter from War Department

French Decoration

April 15, 1920

WAR DEPARTMENT
THE ADJUTANT GENERAL'S OFFICE
WASHINGTON

IN REPLY
REFER TO

AG 201 Tarnowsky, George de
(Misc Div) WSB:EMM:AG B&M Sect.

April 15, 1920.

From: The Adjutant General of the Army
To: Colonel George de Tarnowsky
2719 Pine Grove Ave., Chicago, Ill.
Subject: French Decoration.

1. There is forwarded herewith by registered mail a decoration of the Order of University Palms, grade of Officer d'Academie – Silver Palms, awarded to you by the French Government.
2. It is requested that the inclosed receipt blank be executed and returned to this office.

By order of the Secretary of War:

Adjutant General.

2 incls.

Post War/Post Armistice

Mission 1919

A Sally into Serbia

Curated from A Sally into Serbia

January 24 – February 11, 1919

Again our luggage handled by prisoners of war looking like brigands, and we watched particularly our food, while waiting for Col. T to find A.R.C. headquarters...In these few days we had seen much novel and instructive and had passed through several distinct racial and political regions.

Long Tunnel into Italy

Friday, January 24th, after a very busy day of final preparations, I left Paris at 8:50 PM, one of a party of nine. On the morning of January 25th when I awakened, we were at Aix-les-Bains, and the chillness of the mountain air quite accounted for the marked cold I had acquired. I got up for coffee just too late at Chambery. The snow on the mountain sides here was the first I had seen so far in the winter. We followed up a narrow valley with vistas of the Alps on either side until we reached Modane about noon. Here, we had two hours to go through the customs and Liet. Holmes, Infantry, of Denver, R.T.O. there, led us to a village hotel where we had an ample meal, and then we again boarded our train. An electric locomotive had been attached at this place and immediately starting after a shorter tunnel, we plunged through the long tunnel into Italy that was one of the feats of hand-work, for this work antidated the use of machine drills in tunneling. Emerging on the southern side of the Alps, the change was immediately marked. The houses differed from the French in being more rectangular and, like our Colonial architecture, in proportion. There, Carabinieri in their Hapolsonic hats were on the platforms, and the absence of the French poilu and the varieties of Italian uniforms also showed the change. The railroad was electrified, and we passed down a valley with a wonderful background of snow-clad Alps on either side, filling all preconceived ideas of what mountains should be. The narrow flat-bottomed valley was highly cultivated, and terraced sloped vineyards ran high up on the steep sides. The houses were not as in France - congregated sharply into villages, but were frequently isolated, and division fences or lines plainly marked off the land.

Italian Barbiere

We gradually approached the floor of the valley as it widened, and after dusk, ran into Torine - most suggestive of a U.S. town in this half-light, with modern factory buildings, electric lights everywhere, trolley cars, and the large station where we stopped crowded with people with alert looks and energetic movements. Having some time here, we went out. Mason and I, both feeling so somewhat the worse for wear from our cold, found a little bar where two glasses of a luxury - hot milk, water, and sugar increased our spirits perceptibly. Down a wide street with a central parkway, past modern buildings, we wandered. People turned to see us. Evidently, Americans were not common, and in a shop, we experienced the hand work of an Italian barbiere for the first time. Sitting upright and having the head forcibly thrust back at an acute angle, in order to allow the wielder of the rasor to perform properly, was a novel experience. On the way back, we bought a map to orient ourselves through Italy and reached the station. The streets are straight, wide, modern in aspect. Altogether, the aspect of this city of 427,000, at an altitude of 239 meters, was, from this slight glimpse, most attractive. In the station restaurant, we adopted spaghetti as our evening meal. Feeling better, we resumed our places on the train. The trains here are also more American in appearance. The locomotives are large and rather characteristic American lines. Coaches are longer, also quite different in appearance from the very short French coaches.

Weaving through the Italian Alps

After a good night's sleep, I awakened the morning of the 26th at Florence, for we had made a detour from the coast because of a bridge gone out. Florence, 232,000 population, at an altitude of 49 meters, suggested a subsequent visit. We went south from here, along with a high mountain range snow peaked to our east, while flatter plains lay west. The country was neat and looked prosperous, though little traffic was seen on the roads and I saw no live stock for a long time. At noon, we saw some sheep for the first time and later, Roman cattle, quite different from the French and Swiss cattle in their grey color and with their huge horns. As oxen, they offer an obstacle to traffic going the other way. Tuscany was beautiful – vineyards everywhere with a curious appearance due to the fact that clipped trees were used as supports for the vines. Also, we saw cypress trees shaped like Lombardy poplars with their dark color. As we ran along, many hills were topped with towns or strongholds. We had extended views of plains and mountain range. Quite unlike the quick changes occurring in France, the country was more somber, much like the Connecticut valley in regularity. Going through Umbria, the landscape was more desolate, slopes more irregular, broken and eroded as Mason said resembling markedly the Yangtze valley in China. It was less attractive – apparently much less thickly peopled.

Distinctively Rome

Through a winding valley, we followed the Tiber into the plains. We saw the dome of St. Peters afar soon after running into Rome at 2 p.m. The station here is large, but without the bustle of Torino. Going across the street, we found no rooms available at the nearest hotel, so by advice went by caressa to the Hotel Regina. However, Mason and I changed very soon to Hotel Flora, which was in the same building as A.R.C. Headquarters. We then stretched our legs by walking through the Borghese Park. This offered wonderful contrasts in foliage and vistas. Cyprus and myrtle trees abounded, old statuary decorated unexpected places, and remnants of older buildings also appeared. Our wanderings brought us back to Piazza Populo, with it's huge gateway, into the large circle in the valley with the obelisk in the center. Overlooked by the Borghese hill, the twin churches on the opposite side giving a characteristic difference from anything we had seen. We wandered through narrow streets, which are peculiar in being paved with quite small granite blocks sat diagonally, and eventually reached the hotel. The architecture of Rome is straight lines, heavily corniced roofs, and windows, absence of the ricoco curves of the French and flat roofs were noted. Throughout Italy, the wide views, the detached houses with their straight lines and flat sloped shallow roofs, distinct property lines in the land, villages without manure pile adornment or live stock apartments, showed a people racially distinct, and their appearance bore out the same facts. Many of them have a distinctly Semitic cast.

I found that Col. T had left the previous night for Bologna, as I had missed him again. The hotel at night, where we were en pensione, gave us a good dinner and later seemed quite giddy with women in evening toilettes and much chatter, music, and sociability. Rome gives the impression of being a city with a distinct social season, and evidently the winter sees it at it's height.

January 27th - dark and drizzly. After reporting to Col. Anderson of the Croce Rosa Americans, I found a party was to go out that night for Serbia and that Col. T was to meet it at Bologna. Though feeling grippy and with a cough of annoying spasmodicity, I decided to go along in order to overtake the Colonel and to acquire some information about what the future held, as far as plans were concerned. After straightening out the necessary business here, I went over to the American Embassy with Mason, who had a letter to Ambassador Page. After lunch, like in Paris without butter, though milk was obtainable, we spent two hours driving about the city in a caressa, which is the characteristic vehicle of Rome, antiquated in appearance with small wheels in front and a huge umbrella to be hoisted over the driver's head in bad weather, and with very small horses that seem, however, surprisingly efficient. They offer excellent sight seeing opportunity. Our driver spoke English haltingly - surprisingly well, if his statement that he had been studying it only three weeks was true. We drove through the park again with different vistas, past the old city wall, across the Tiber, about the size of the Chicago River and not especially impressive, past Adrian's tomb to St. Peters. St. Peters, facing a large piazza and paved with this characteristic small diamond shaped pavement, is flanked on either side by a colonnade forming a semi-circle. The marble is dingy, yellow, and markedly weather-worn. Statues, worn by time, crown it's parapet. The steps approaching St. Peters are almost like a ramp, the slope being such that one of our icy sleets would cause marked havoc to pedestrians over it. The Vatican to the right is also yellow and, with small windows, has a distinctively impressive exterior. Going in for a few moments to St. Peters, I was startled. It was so different from all Gothic architecture. No stained glass, but the walls fairly vibrated with color. There was much heavy gold, and the altar was a mass of huge golden ornaments. The general impression was rather of barbaric ornament in it's blazonry, to one accustomed to the subdued traceries of the English and French cathedrals.

En Route Scenery of War Territory

We had dinner at the station and started with Major Woolston and Captain Nevin, A.R.C., Miss Corcoran, Mrs. Phillips, Miss McLane, Captain Fox, Davis, Lieutenant Voor, and Cornwall. We again were fortunate in having a wagonlit through to Trieste, and in the same car were five Serbians returning with passes visaed by the Minister of War, one the President of the Serbian Red Cross, another Captain Yanovitch, Secretary of the Serbian Red Cross. The former spoke only French, but Capt. Yanovitch spoke a hesitant English, as well as Maj. Woolston, and I had some hours of conversation during the evening. Also, in the car was one of the Royal Family, Duke d'Aosta, a Captain in the Italian Army, and an accompanying General, also bound for Trieste. We left Rome at 7:35 and had a comfortable night.

The morning of January 28th came dark and lowering. At 5 a.m. when we passed through Bologna, no Col. T was to be seen. Through Padova about 7 a.m., through Mestre where the station shows the effects of bombing raids that resulted in a number of soldiers being killed, and then to my surprise, we ran into Venice about 9 a.m. - through lands flat and largely covered with water. Across these, a huge dirigible shed could be seen. We looked about for the missing Colonel here again without avail, walked out to the front of the station, saw the Grand Canal, and then reimbarked. From here on again through Mestre, we proceeded through lands as flat as about Chicago. In the background to the north ran the wonderful Alps with their serrated snow-covered peaks. Villages could be seen at a great distance across this plain, especially from the high square towers that seem characteristic here. At Treviso, we saw our first signs of war territory, though it suffered largely from bombing alone. There were barbed wire entanglements, however, trenches, shell-shot houses, and soon we crossed the plate Piave, famous in the great Italian retreat. The banks are low - perhaps 6 or 9 feet high, with the precipitous appearance of Missouri River and banks. The bridges across are necessarily long ones and were of stone arch construction. Many of these arches had been blown up, and temporary timber work enabled us to cross these gaps. On the highway bridge above us, convoys were crossing on a corduroy road over the sand banks with small bridges of piles over the water courses. On through Conegliano, Sacile, Pordetone, Casarca, we passed to the Tagliamento River, wider than the Piave, but with the same sandy bed and narrow water channel. Very little water passes down it at present. The same long bridges are thoroughly blown out at places, and our progress across the temporary repairs was very slow. Everywhere though this territory were many Italian troops, and the civilian activity was not apparent, few being seen. We saw some of our own dough boys here and found that some battalions have been in this district. Through Codicipe and Udine we went. They show marked destruction, though nothing like the razed condition of the Verdun sector. At Cormens, we passed the old frontier and went on to Corizia. Before reaching this place, there were very sharp hills thickly covered with trenches, and we crossed the Leonze River that runs through a narrow valley, which shows much evidence of warfare. Corizia shows surprisingly slight damage, being spared by the Italians in their offensive. Because of the Italian character of the population, what damage was present was usually from Austrian fire. We went through the Corso, huge hills where the Italians first experienced gas attacks that caused great mortality. Fierce fighting occurred here for weeks, and the Duke d'Aosta pointed out some of the interesting places as we passed them.

il Boore

Dusk began to fall as we reached Trieste about 6:45, running down a long hill, through a tunnel into the station that gave a marked appearance of emptiness, for only soldiers were about. Sentries were on duty at all exits, and it was difficult to get porters. Finally, we managed to get our luggage outside and on to a small cart, man-pushed, to take it to the hotel. Capt. Robinson, A.R.C., met us and led the way on foot to Hotel Savoia, which he said was a few minutes away, though it proved to be a very trying interval, for a terrific cold wind was pouring down from the mountains, known as il Boore. This was about 60 miles an hour, and cape had to be held on by hand. The streets here are paved with large flat stone slabs. The architecture is heavy, severe, but without the cornices of Rome and not as attractive. We went along the waterfront before reaching the hotel, and some of the gusts nearly blew us into the sea. The quays here are built out at right angles to this front street and have, at times, chains along them because of this fierce wind. There were very few people on the streets, the iron shutters were all down before the shop windows, and the abundance of military activity showed how near the frontier we were.

Hotel Savoia

Hotel Savoia was formerly Austrian, but now is Italian, the furniture showed the Neue Art of Austria, Bentwood chairs and golden oak tables, beds with a small square comforter over the lower half only. The Hotel was brilliantly lighted, jammed with people, and was a welcome change from our railroad trip. The money used here is kronen most largely, the exchange varying from 200 to 250 kronen for 100 lira. A table d'Hotel dinner for 22 kronen was good, though the bread was apparently the most expensive article on the menu. We heard German spoken here at times by waiters and the hotel maids. Altogether, there was a most curious admixture of racial types and language. We soon were found by Col. McIntosh, G. M. C., who heard that nurses were to arrive with us, and he took our two nurses immediately after dinner up to a private sanitarium to take charge of Lieut. Wood, U.S.N., who had just been operated on for gangrenous appendicitis. Although there is a considerable naval and military personnel along this region of the Adriatic, I was told that there was not a single hospital nurse assigned to that district, and they pounced upon our two nurses with avidity. Later I met Commander Van Auken of the U.S. Destroyer Stribling, Lieut. Wood, C.O., and Commander Pence, U.S.N., Port Officer. I was much impressed by the high type of the naval officers so represented by their courteous demeanor and interest in helping strangers from the States in every way and their power.

On January 29th I awoke feeling rather grippy, for the heatless rooms and the heavy stone construction was not conducive to improvement. The high wind continued without sun.

Attaché Armstrong and his Bride

The next morning, Jan. 30th, il Boore was still blowing unabated. After breakfast in our room, the Colonel and I went to the private hospital with Commander Pence to see Lieut. Wood. We started in a Fiat car but finished on foot, because of tire trouble. Later Col. T and I came back in a side car to the military headquarters, then went to the commissary and stocked up our food boxes with corn beef, hash, coffee, cornwilly, grape jam, candy, canned soup, and then to the market for bread and oranges. Later, we packed this food in a large box to accompany us. At the hotel, we had some very attractive white buns, but these were at a distinct advance in price. We took the luggage to the station about 4 and boarded the train at 4:30 to secure seats. There was one out of the long train with one first-class compartment in it, and I had our luggage arranged when the station master appeared. With frantic gesticulations and voluminous words, he hoisted me out of it, as it had been especially reserved for a Lieut. Herbert Armstrong, Infantry, and his bride, Helen Byrne, who were going to Belgrade with 17 trunks, he to be assistant military attaché there. As this was the only decent place on the train, I felt somewhat aggrieved, but saw no recourse. The only thing left was an adjoining compartment, where there were two sections of six seats each in which one had a seat complete, two others had sub-cushions minus the seat, while the other 9 exhibited only the frame work where the seat ought to be, a vacancy remaining between the woodwork. By filling three of these in with our blanket rolls for seats, however, we were quite comfortable. Fortunately, I left my four heavy pieces of luggage at Trieste on advice of Col. T.

Italian, Jugo-Slav, or Austrian

The locomotives now were wholly unlike those through Italy, distinctly complicated and resembling Austrian railway stock. The cars were battered, many windows being boarded up, glass being gone, and evidently, we were near the dropping off place. The train was packed almost entirely with soldiers, our compartment filled up with officers. As we climbed up the long hill past Miramar, a summer villa suburb, we reached a junction Nabresina, from where we started back from the Adriatic up through the mountains. The grade was evidently quite stiff, snow appeared on the ground, and darkness fell. Soon, the only light was the glowing tips of cigarettes being smoked by the Italian officers, and their singing helped to pass an otherwise lonely time. We passed, during the next hours, several trains with Hoover provisions, guarded by dough boys, on their way to Vienna. Finally, wrapped in blankets to keep out the increasing cold, we reached, at 11:30, the frontier station Longatico, Logates, or Leitsch, according to Italian, Jugo-Slav, or Austrian vernacular. The bitterness between Jugo-Slavs and the Italians was evident, manifest in the talk at Trieste. No information could be obtained as to what lay beyond this frontier post at Trieste - the difficulties of interchange of money, of travel, and of information, evidently a concerted effort to inhibit intercourse. The Jugo-Slavs accuse the Italians of padding their population figures. Italy hates the Jugo-Slavs, who have taken advantage of Austria's sudden armistice to overrun much territory. This frontier post represents the region where each country, seeking to gain as much ground as possible from that formerly ruled by Austria, came almost to an open clash. In view of the [blank] manifested toward each other, President Wilson's present marked popularity with each faction seems a rather difficult position to maintain. Half asleep as we were at this time, we were dumped out into snow, the first we had seen since Modane, and had to light candles to get out luggage pulled off from the old car. After a wait, we got more soldiers to help carry our heavy stuff into the station some distance forward. Armed Italian guards were on sentry duty everywhere. A search of this town revealed no hotel accommodations available. We were told the train on which we would proceed would be in at 4 a.m. by which was two hours too optimistic. In a large room, the custom officials went through all the baggage of the passengers, scrutinizing the papers with care, though as an Americano Missione, we escaped with but a minor survey of papers. We eased matters by a round of hot coffee, made on the Primus stove, with some bread and jam. Then the train we arrived on became available, the other four in the party went aboard with blankets to stay and await the expected train. However, they found the unheated car and the darkness very trying and had little comfort during the ensuing hours, even though wrapped with numerous blankets. I stayed in the station to guard the luggage, particularly the food box. I was warm and slept several hours. One man in civilian clothes and cap was prominent in this room, looking very like a mid-western yard railroad or switchman in appearance, the largest man of the men there - apparently an Italian secret service man. I woke up once to find the room jammed with Italian soldiers, and again to find two U.S. soldiers from Cleveland,

Ohio, standing near, in from a Hoover train. I had quite a chat with them. They had been at Fiume and were not enthusiastic about the Italians. At 6 our shuttle train appeared, and we secured a compartment therein, again had hot food by the aid of candle lights, wrapped ourselves in blankets.

Americana Missione

Through the dawning light, I watched our progress as we wound down from the mountains with heavy pine growth somberly lining each side, until we reached a broad flat valley and followed that into Ljubjana, or Leibach. The trainmen on this trip wore the curious Austrian caps and blue uniforms seen in old pictures. Everywhere, Serb soldiers were also on guard, and the military control of this country was apparently Serbian. We simply had to show our papers, say "Americana Missione", and all was smooth for us. We were apparently remarkably welcome. At Ljubjana which we reached about 8 a.m., we left our luggage in a very modern station and were escorted by a soldier to the commanding officer's headquarters, where rooms were requisitioned for us at the Union Hotel, although over the telephone, they had said they had none, and we walked on there. A Serb captain came up, introduced himself and assisted in directing us, speaking English very well. He said we were the first he had met since leaving the States in 1914. He came from Iowa. At the hotel, which had a driveway separating the two portions of the ground floor, we had to wait while rooms were cleared of occupants and cleaned for us. We had coffee and rolls, quite German in appearance and taste, with sugar also. Then with an opportunity to clean up and sleep, we separated. I found that the leaving behind of my haversack for coffee had cost me a box of American cigars. The hotel proprietor exhibited much evident anxiety when I reported this. He sent out and bought a box for me, so I said no more, although the new lot were a fearful exhibit. Miss Burke, only one of the nurses, went down for lunch. We had very good meat and potatoes, with a light Hungarian wine, and then real russet apples with flavor and texture not seen since we left the States. Kroner was the money in use and German the language. The horses here looked distinctly good in quality and well kept. I saw for the first time a new hitch, which consisted of a single horse along the ordinary wagon tongue, so that it gave a curiously incomplete appearance to the vehicle. Walking through the town, we found it modern and German, hence uninteresting, shoddy, and in poor taste. An aeroplane flew above during this walk, and the shop windows had only cargoes of cheap gew-gaws, the clothes of ancient pattern and distinctly inferior quality. The people on the street wore noticeably old and worn clothes.

The Armstrongs Dashing Ahead

I'm paying our respects to the Serbian General of the district, whose headquarters was in the hotel where we billeted, we encountered the Armstrongs again. He apparently had just been decorated, for now on his tunic was a huge bit of junk, which quite outshone his previous sole decoration of "pistol expert." The Armstrongs seemed quite ill-adapted to represent the States so far away. In marked contrast to the naval officers we met, they were not only snobs, but apparently proud of it. The anxiety they showed in dashing ahead at every stop in apparent anxiety lest only one room should be available was noticeable, and especially in the ride from Trieste, where they had several empty well-cushioned seats, they showed no inclination to extend the slightest consideration to our nurses. This rebounded on their heads later, because our cooking facilities were not shared with them. I am afraid that their gullets were chilled by the absence of hot tea and coffee we were enjoying.

"Je suis jereste!"

On boarding this train, we found the best accommodations we had between Trieste and Belgrade. An upholstered, gas lighted, and steam heated compartment with table especially reserved for us now. The Primus and the food box gave us a good hot dinner of soup, corned beef hash, bread and jam, and coffee, after which we played bridge for several hours. Meanwhile outside, we had a regular pyrotechnic display. Our locomotive was a wood burner, with a smoke stack of the tremendous size now obsolete in the States. The pine produced sparks in such amounts that they made a fairyland outside the windows, as sheets of them blew past the windows and fell on the snow. We also made better time, with much less straining, struggling, and panting of the engine, with only slowly revolving wheels, for snow was falling and the night was dark on our return. Some of this was seen to be through narrow valleys of very steep slopes, where the town was strung along, with only room for a single street. At 11:30 p.m. we reached our next point of change, Agras, or Zagreb. We got off in a very good well lighted station, hunted up the military command again, who gave us requisitions for 3 rooms in as many hotels. There were no horse-drawn vehicles about, and after some delay, the aid of the C.O. secured two-wheeled cart manned by two men. The town seemed some distance away, and after conversation with us in German, directions were carefully given by these men in Croatian, for Zagreb is the capital of Croatia. This was repeated several times, and our luggage was piled high upon the small push-cart.

Apparently, we were ready to start, when as many more persons appeared with additional luggage. In the gloom, theirs was added to ours. Of course, the cart rebelled and tipped it off helter-skelter. I felt worried, more especially for the food box, but we finally got all of it separated and loaded again, and off we started through the snow. One Croat between the handles pulling, one behind pushing and trying to keep miscellaneous luggage from spilling off. They only negotiated this load by keeping the two wheels in the trolley track that we followed. Everything was dark and forbidding. The language was strange, the buildings shuttered, and we picked our way along carrying haversacks, canteens, and small bags. We looked much like a covey of anxious turkeys spread out over the icy road. We reached the first hotel where the Colonel was billeted. This proved, on trial, to be high in price, as well as in flavor, for it was extremely dirty. He had his bill raised because of his rank, adding insult to injury. We next left the three nurses at a small building, where a vigorous ringing had at last opened a door of a tunnel-like passage into a courtyard. Two beds and a couch in a room gave them shelter and a chance to lie down here. The rest of us proceeded on. Gradually, we dropped the others until I reached the Palais Hotel, where my requisition called for a room with only one companion left who spoke French slightly, as well as Croatian. We rang up the porter and entered a gloomy high hall with stone steps leading up to the office floor and a stairway to the upper stories beyond. I gave the porter my requisition. He threw up his hands. He spoke only Croat and German, and that was beyond my ability, but remembering the words of my former Bunkie, Maj. Sam Elsworth, I also said "Je suis jereste!" and dragging my baggage up into the office, for the bulk of this had been left for me to take care of, I found there a good fire going in a small stove, while around on chairs with heads resting on their arms were several other men asleep, evidently also up against over-crowded hotels. I, however, arranged my blanket rolls like a seat with a back rest, calmly sat down with my feet close to the fire and looked at my watch. It was about 2 a.m. and February 1st. The Armstrongs had arrived here earlier and presumably had gotten a room by virtue of their diplomatic relations. I slept very well until 6 when the day's work began to stir in the office. The Palais Hotel is fire proof, having very heavy masonry walls and tiled floors, but the attractiveness within was not marked to me at this time. At 7, I found a cafe, and by the aid of a hotel woman who spoke some English, I managed to get a good cup of coffee with bread, and an unexpected addition of ham and eggs. This put me once more again in the running.

Armstrongs with their Innumerable Trunks

After this again by a man-drawn cart, I took the luggage to the station of departure, a different one from where we arrived, and not far from the hotel this time. Down a street with a trolley line on it and lined with really imposing buildings which faced a parkway in which were occasional monumented buildings evidently of a public nature. It was snowing and bleak when I reached the station at 7:45 a.m. for the 8:30 train and waited for the others. The Armstrongs arrived by horse-drawn carriage soon, their innumerable trunks also in one of the vehicles made familiar in pictures of Russian war transport wagons. These are small carts on four small wheels with a crate-like frame work, both holding about as much as one of our dump carts. From the top of this frame work supports came to the ends of the axles outside of small wheels that are perhaps three feet in diameter. The horses are small, and the axles are about 7 feet apart. The whole vehicle has a very crude appearance and looks as though much of it was merely of natural shaped branches of wood. A Serbian officer was with the Armstrongs, evidently to assist their linguistic difficulties. The rest of our party soon came on sleigh, reporting that the Colonel had a narrow escape from being devoured by an irate dog that fastened his teeth in the heel of his shoe unabatingly. I had been mistaken frequently before, thinking that I saw them approaching on foot, remembering the night before, for great numbers were pouring into the station from every direction on foot, all carrying bundles and such a mess! All sorts of combinations. Peasants in native costumes, Croats, gypsies, soldiers in Serb and Magyar uniform, for the trainmen wore the latter. Clothes of every vintage since the Napoleonic wars were seen. The foot wear was very bad, oftentimes old boots and shoes, or no boots or shoes. The mass of ill-kempt humanity jammed together for this one train eastward in the 24 hours.

Serb Diplomatic Mission

Through Col. T's linguistic ability and potency of the Amerikansky Missione, we again got very good accommodations of what was available, being given two sections in a special car, the military attaché et femme having one in front and also two Serb soldiers to guard the car. Our train, meanwhile, stood for two hours over the schedule time to allow an important Servian diplomatic mission coming from another direction to be attached to this one train. In this interim, we stowed our goods and ourselves in the first-class waiting room, of which the doors of entrance and exit were not only locked, but were guarded by Serb soldiers with bayoneted rifles. It was warm however, and there were some big round brass topped tables and comfortable chairs therein, so the Primus was started and breakfast was served. We found this manouver cause much curiosity and observation. Evidently, we were strange beings, a strange equipment for cooking, and we were as strange to them as they undoubtedly were to us. When we got aboard, we found our home for the next 36 hours, as it proved, not elegant – windows in the corridor broken out, doors to our compartments in rickety condition, seats unpadded were covered with a paper fiber cloth like gunny sacking. There were no lights and little heat. Of the toilet room, it is best to say little, save that it was the limit. On this train were 35 coaches. Our special car just behind the one of the Serb Diplomatic Mission, one behind was filled with French officers bound for Belgrade. One far behind with three British Tommies taking up a month's rations to Belgrade from Fiume that T. succeeded in having put on the train, to the great relief of the Scotch sergeant in charge.

Sava River

As for the other cars, they simply exuded humanity. Never did one see more huge bundles. A most motley throng all carrying food, usually apparent by a sticking out somewhere of a long loaf of bread. Finally, we started slowly out, soon crossing the Sava River, here about the size of the Des Plaines at Joliet, running through a flat valley with earthen banks about 8 feet high. We followed this River to Semlin, it's flatness was that of the Kankakee region. Only at one place did the hills come within view. On the map, the line runs southeast for a long distance from Agram almost straight, then almost east. This angle is where the mountains came down within close proximity. All afternoon, we rolled slowly along. Outside a regular blizzard, hard at it. Our progress was such that we speculated on breakdowns or hold ups from drifts.

Mob of Passengers

At times, our waits at stations seemed interminable, but eventually on, we would crawl. Some heat came from the engine during the day. We had lunch. Between us, one compartment for coffee and the other for drying room, but with darkness came cold, so our candle elimination was soon extinguished while we stretched out on our seats for the night, the three nurses in one compartment, and T and I in the other. I slept quite well until about 4 a.m., our guards also being asleep, when at a junction point, a mob of passengers suddenly descended on us, overjoyed to find so much apparent available room. The nurses were greatly startled by this invasion as were we also awakened suddenly. T shouted "Americanish Missione verboten", dove into the corridor in his slippers, and with the guards finally awakened, the horde was cleared out. We again continued our repose.

Irridenta Region of Styria

Several times, we stopped for a long period during the night, the engine apparently going off for fuel, and this was apparently at times wood, for we had again the spectacular sparks illuminating the night. The trouble apparently was in the fuel. Only "brown" coal was available and wood. The Italians controlled all the good coal in Styria. Question: Is this economic factor any reason for the intensity of the Italian demand for the irridenta region? Plenty of rolling stock for more trains was seen, but there was scant motive power, many engines along the road and at Semlin being out of commission, because no repairs were possible from lack of material and skilled workmen. A trainman spoke feelingly to us, at one station, of his having been a prisoner of war for four years in Austria. He had lived in Fiume, but he said that it was certain from late developments that his imprisonment had been originally due to Italian accusations.

Roast Beef and Potatoes in Semlin

At 8 a.m. February 2nd, we had breakfast. All day we progressed through this flat, uninteresting country, which, however, looked very fertile, well tilled, and thickly peopled. We saw carts drawn by oxen and very good looking horses - also in some woodlands, droves of hogs. The snow continued throughout the day, and it grew colder. We finally had our entire store of blankets about us and craved more. Cold hands, cold feet, draughts. Our sole respite was frequent hot drinks, coffee, soup, tea, and food. About every three hours, we would stoke up our calories in this way. We finally reached Semlin about 7 p.m., only eight hours behind our very slow schedule, but relieved not to be left in a snow bank. When we piled out with our luggage, we landed in snow drifts with a bitter wind driving past the train, which caused us to hurry into the station as fast as possible. Here, the military commandant was very courteous, gave us a requisition for rooms in the town about a mile distant, and also a carriage to take us there. Arrived at the hotel, we found a French soldier standing guard at the entrance. Here we had a good hot meal with the best roast beef and potatoes since the States to comfort us, and then turned in for the night. I went about 400 yards onto a second hotel, mistakenly called the Grand. We had to stay over night here, because the last ferry for Belgrade left at 5:30 p.m. My hotel room was chill and damp, the stairs going up from an outside carriageway, but I slept with only my outer clothes on, and my shoes replaced by wool lined slippers, using the lining of my rain proof as an additional protection under the bed clothes, with the overcoat itself outside.

Ferrying Down the Danube

February 3rd. After bread and coffee en lit, I dressed in this chill clammy room and then went to the Central Hotel, where I had a second breakfast of soft boiled eggs. After which, I brought the nurses down and directed their effort to the same end. As they were eating, the Colonel appeared. We assembled the luggage on a sleigh, and I went with it to the ferry. Luckily, the high wind had abated, though some snow still fell. The day was dark and grey. Here we saw many of the small wagons described before. Many used the single hitch described at Ljubjana and drawn by a small horse hardly larger than a donkey. At the ferry, I waited for the rest of the party. The Danube was about the width of the Hudson River at Albany, with low banks of mud, but it is about 10 to 16 feet deep. We crossed by gang planks onto steel barges anchored along shore as wharfs. The British Tommies had already gotten their stuff on these lighters with Bulgar prisoners of war as porters, and they had them handle our luggage also. It was like the Orient here. Pass, Tarbouches, fur hats of the Tartar shape, old clothes, an overcoat made out of gunny sacking. Peasant women and men, the former with short skirts hand embroidered with short leather jackets embroidered at times and lined with wool. Picturesque, but dirty. All material handled by porterage. Pandemonium reigning. Presently, the ferry came along, a side wheel craft, narrow, long, with the lines of a small English channel boat. It was also packed with a motley crowd, who poured off as soon as they docked. We saw many poilus, for the French are now routing troops home from Saloniki by this route. Increasing numbers of Serbian uniforms often obviously added to from French or English sources, with the distinctive cap of the officers, which resembled somewhat our overseas cap made with expanded top, stiff sides, and with a dinky little visor about an inch wide coming sharply down over the forehead. Most striking where the huge shoulder straps superimposed on the tunic like our marine corps wear. Our Tommies were very appreciative of cigars. "Now we'll have something in front of our noses," they said. We pushed off, went down the drab Danube a mile or more, then up past broad flat islands into the Sava river to Belgrade. This city lies on quite a high promontory between the two rivers, and at the tip is an old Turkish fort picturesque and showing abundant effects of Austrian bombardment. The modern fort lay below on the lower land. In the river was quite an amount of shipping, comprising steel barges, monitors, and gun boats, some of these British, side wheel steamers of long narrow lines. The quay was crowded with a fleet of barges loaded with Serbian cavalry, and we had to wait until four of these side by side were hauled out by a tug. As we approached, the Oriental nature of this waterfront was apparent in the clamor; dirt, variety of costume, and when our luggage was put on the adjacent lighter, innumerable claims of assistance were entered by loiterers about. We saw here a peculiar wooden rack for porterage that rested on the back, hedging from the shoulders and terminated back of the hip by semi-circular bolster. On this device was carried huge boxes, casks of wine or beer, tremendous loads for one man and especially wonderful, considering the slippery wet footing. Many wore the native shoe, which are like our Indian moccasin as far as the soul of the shoe, while the upper part is formed by the plaiting of the split sides of the sole, and there is a turned up tip like the familiar Turkish slippers for a while up the leg goes a circling of straps. Inside was worn plenty of cloth, but in this wet slush, the combination did not look attractive.

Ashore to A.R.C. Headquarters

We crowded ashore. Again, our luggage handled by prisoners of war looking like brigands, and we watched, particularly our food, while waiting for Col. T to find A.R.C. headquarters from where he sent down a Ford camionette driven by a Serb with an A.R.C. Lieutenant. Our dunnage was loaded on this, and thus freed, we walked to headquarters. The street along the water front was crowded. Very few horses, vehicles, or motors, but many ox carts, soldiers. Debris of every sort. Above this lay the main town. To reach it, we climbed quite a considerable flight of stone steps slippery with snow and ice, a shorter way than the vehicles went. On our left, after we reached the top of the hill, was the Cathedral with a tower distinctly of the Greek church type. Headquarters was not far, and we were warmly welcomed here. It was the ground floor of a modern four story masonry building, attractive in appearance, with a courtyard beyond. Just across the street was a huge brick building apparently left untouched for months, after having gone just above the first floor in construction. Being noon, we went to the Grand Hotel for lunch. Most of the A.R.C. forces, now about 35 in number, eat here, though they are billeted in lodgings throughout the city, as the hotels are handicapped by having had all their furniture removed by the enemy during their late occupation. We entered the hotel through a driveway into a court from which we turned into a large high ceilinged room about 25x60 feet in size with one and filled with small round metal tables for coffee or drinks. Beyond them was a billiard table or two, and in the far end was a bar scene so commonly through this southeastern country. Through this room, we went into an adjacent room lighted only by skylights, about 35x50 feet in size, where at night, an orchestra played, and small tables were available for either coffee or as an overflow from the main dining room. Only beyond this did we find the regular dining room. This was about the size of the first room, very ill-lighted both by day and night. The napery showed signs of much previous use since seeing the laundry, but the food was abundant, and the prices not comparatively excessive. The menu was in Serbian with it's unfamiliar alphabet and in French. While the prices were quoted in Dinari, but Kronen were used largely in actual payment. Kronen to be accepted here had to have been stamped by the Serbian Government. The exchange was from 330 to 345 Kronen per 100 Francs. The Lira was practically unchangeable here. The price in Dinari multiplied by 2 1/2 was the price in Kronen. Our meals averaged about 22 Kronen. Some sugar was available, but no coffee was to be had. The so-called white coffee, or "orsatz", took it's place. The most striking thing, however, was the great amount and variety of fancy cakes seen in all shop windows. The gateaux in prefusion and taste exceeded by far the best Paris could offer. Bread was curiously very scarce and was charged for by the slice - about one Kroner for each, or 5₡ of our money. We saw many ill-clad people on the streets. The shoes, however, seemed distinctly to be the most needed thing. Fuel was scanty. For the personnel, the A.R.C. was distributing a little wood by it's camionette.

Billets in Belgrade

After lunch, we went to our billets. It was still snowing, and as our luggage had gone on by motor, we walked out. I was billeted where Maj. Woolston and Capt. Nevin were located, about a mile from headquarters on Ulica. Capt. Yanovitch also was temporarily there. They had arrived Saturday in time for the evening ferry. Capt. Yanovitch, on looking up his home which he had not seen for four years, found his previous furnishings gone entirely, with the exception of two arm chairs. Our billet was a very attractive one. To reach it, we went out a sort of boulevard, past the old palaces of the King and a new palace not yet completely finished of the Prince Regent, the latter quite attractive in structure and design. Above this, in a building since torn down, was where King Peter and his Consort, Queen Draga, were killed. The Prince Regent was living across the boulevard in a one story stone house of attractive exterior, rather French in appearance with the entry from a courtyard, which was guarded by two Serbian sentries. Our quarters were the second floor of a two apartment house of distinctly solid construction, cement stairway, concrete walls, and floor covered with tile and with modern attractive furnishings. Two bedrooms, each with twin wooden beds, gave us comfortable resting places. We had, in addition, a sitting room heated by a huge porcelain stove, which ran from the floor almost to the top of the very high ceiling in which a small wood fire really produced very effective results. Behind this was a dining room, kitchen, and an unexpected luxury, a bathroom, with facilities for hot water.

Conferences about the Mission

February 4th was also marked by a heavy fall of snow. We were told that until a week previous, no cold weather or snow had occurred, but during our stay, the conditions were very wintery. Sleighs were used, and everyone bundled up markedly. This second day I did little, save exploring my way from headquarters to billet. Col. T was billeted with the A.R.C. Commissioner for Northern Serbia, Col. Farnham who had with him also several of his staff, having a house with a staff of servants and so having very pleasant surroundings. Conferences took place between Col. T., Col. Farnham, Major Woolston, Maj. Perkins, and others, regarding the possibility and scope of the work to be undertaken. The great difficulty lay in the problem of transportation. We had experienced this coming in from the west. From the east, supplies were brought up by the Danube by boat. Should it freeze, serious interference would result, and even without this, a scarcity of fuel made the traffic much diminished. From Belgrade to southern Serbia was a waste. The retreating Austrian Army had blown up every bridge and tunnel from Duskap. As both tunnels and bridges were very numerous through this region, the time for repairs would necessarily be very protracted. The road accompanying the railroad had also been dynamited freely. We were told that a camion train of British motors had left Salonika for Belgrade numbering about 75 trucks. 5 had reached Belgrade, the others having broken down under the insuperable stresses and strains of the trip. Food seemed abundant in Belgrade, but this was explainable by the fact that there were no food restrictions, hence the supply apparently abundant might be absolutely terminated unexpectedly. Again, immediately across the Sava lay former Austrian territory. Through this territory apparently food was quite abundant throughout the war, for this portion of Austria, hostile to the rulers in power and in a fertile region, concealed their produce and refused to cooperate with the demands of parts of the empire in dire need, so that Austria was evidently splotchy as to food difficulties. It seemed rather incongruous to see, as we had, people clad in rags sitting in shops consuming cakes, costing 15₵ each with apparent gusto, and taking a considerable number of them.

Col. Sonnemeyer

During my stay in Belgrade, I went one day to the military hospital where Col. Sonnemeyer is in charge. He designed and had charge of the construction of this hospital before the war and has had military charge of it during Serbian control. An elderly man of charm and manner, speaking French fluently, but electing to practice in the English tongue. I heard his twisted words with great sympathy, realizing how my attempts in foreign language appear to my auditors. This hospital was distinctly commendable, of fireproof construction, with tile floors. It was neat in appearance, well maintained kitchen with modern huge steam cauldrons was far from being out of date. We saw about 60 patients here with typhus or spotted fever. These were Czech soldiers who had been prisoners of war in Bulgarian hands and who were thrown back into Serbia in a deplorable condition. A threatened epidemic, as evidenced by the outbreak of which these men were a part, was being vigorously combatted by delousing measures by the Serbian officials. Col. Sonnemeyer, during our stay, had two nephews arrive from Russia after three months of desperate privation and difficult walking most of the way from Moscow to the Black Sea on foot coming around through the two Hagusa and then on foot from there to Belgrade - almost knocked out by deprivation and want of food. They were recuperating and happy at reaching friends. Another of the houseguests was an A.R.C. Lieutenant who had driven an ambulance with the Serbian army in it's retreat through Montenegro. His home was in California, and he was a most interesting competent chap. Another day I explored the old Turkish fort before mentioned, with it's heavy masonry with protecting moats, portals guarded by huge iron riveted oaken gates, dungeons lighted by tiny slits that suggested bygone power and bygone evils. In the parade ground within it's walls was where the Austrians hung many Serbians during their occupancy of Belgrade. I was very much impressed by the general racial type we found at Belgrade. At the hotel, the type of individual was one that would pass unnoticed for singularity or unusualness in any American Hotel - distinctly free from any "foreign look." French was the language used outside of their own tongue, though many of the people spoke German also. Austrian occupancy had only ceased about three months before our arrival. The general architecture was attractive. Many one story residences of planning appearance, oriental in having the courtyard within, and doubtless under less forbidding climatic conditions, our general impression would have been even better. Our nurses were billeted in a less attractive house, but their rooms had the walls decorated with Turkish carpets probably of considerable value, and they were not uncomfortable.

Retracing

As the conference developing the fact that the planned for character of the medical work was not possible and as the A.R.C. European Commissioner had expressed a desire that Col. T. take charge of the office at Bologna, we terminated our Belgrade stay and again started back on February 6th, going across the river in the later afternoon to spend the night at Semlin. Here we were all billeted in the hotel that I occupied on our in trip, but had very much better rooms. Evidently the popular resort since during dinner, a gypsy orchestra played queer compositions that are so effective in their hands and throughout, the people was evidenced a distinctly musical appreciation. Our train started at 6:30 in the morning so we dressed by candle light, found our way through dark passages to the street below, and piled our luggage on a sleigh which I accompanied, while another sleigh brought the other four of the party. I thought it was the most side hill road I had ever gone over, for I had to hang on to the side of the sleigh to keep from falling off. I discovered on our arrival at the station that the luggage and my own weight had been too much for the vehicle. My side had collapsed. The driver needed additional remuneration to pay for this. On our trip back, although we again started with two reserved compartments, the crowds of people were so great that it became necessary for us to use only one for our own purposes, save that our bulky luggage was left in the racks in the other. We again had a Serb soldier on guard, but our progress was just as slow, and the corridor outside our compartment was jammed with people. Some of these rode for hours standing and went to sleep leaning against the window. We found it somewhat of a problem to do our cooking and eating and sitting five in a compartment designed to seat six, but all went well with ingenuity. At night, however, it was still more of a problem. We got our blanket rolls from the adjoining compartment, put them in between the seats. Two nurses occupied one long seat, another insisted upon sitting up, Col. T stretched out on the bedding rolls and a part of the other side, and I sat up in the remaining corner until about 2 a.m., when we shifted, and I took his place. We managed to find room where he could continue to lie down. The night was long and, my bed exceedingly uncomfortable so that about 5 o'clock, I varied the procedure by going out into the packed corridor and smoking vigorously for about an hour. Coming in at this time, I discovered Miss Gerard had also passed a rather sleepless night. As the day was breaking, we watched each other with more and more anxiety to see who should dare wake the others first. When we did wake the others at 7, suggesting it was time for breakfast, these comfortably placed people objected strenuously to such an early meal, but our protests were heeded finally. Hot coffee put forgetfulness again in our mind. During this day, we varied our program by bridge, interspersed with the recurring hot drinks again. It was very interesting to see at the station the fighting mob trying to put three people in a space that would only hold two. Actually, it exceeds realization on the part of one not having seen such a crowd. We had to have water brought in for our tea and, at one station, sent our Serb soldier off with three canteens for water. He had to go a considerable distance, the train started unexpectedly, and the result was that neither the three canteens nor the Serb soldier have since appeared. As Col. T and one of the nurses had personally made such an excursion only a couple of hours before, it shows how narrowly we escaped more serious loss.

Introspective Return

We reached Zagreb February 8th about noon, having made about the same time thus far as going out. We debated whether to come back as we had gone, in through capital Ljubjana or to come down through Fiume. The officers at each station told us the other way was the preferable. Finally, however, we fortunately decided to come back as we went and came on that evening to Ljubjana, again in the luxurious heated cushioned train that we found coming down, but this time without any lights. Reaching capital Ljubjana, we were fortunate in making connections with only an hour's wait for the frontier. To our delight at the frontier, which we reached about 6 a.m., Feb 9th, we made close connections for Trieste and with only time for a hot meal in the frontier station. It seemed too good to be true when we reached Trieste by noon, 24 hours quicker than our out trip. Resting up there that night, we came on the following morning, leaving Trieste by U.S. Sub-chaser #234, Mr. Eshom in command, for Venice - a run of about five hours in this way, as opposed to 10 hours by rail. The sun shone brightly. The sea was smooth, because the dreaded il Boore was not in operation.

We hugged the shore in order to avoid the mine fields not yet swept. As we passed the south of the Tagliamento and Piave Rivers, we could see looming up in the bright light towns with high towers apparently emerging from the sea. We reached Venice about noon under ideal conditions and were on shore in time to have an excellent lunch at the Daniele. In the afternoon, we went through the Grand Canal by gondola. Col. T. and Miss Burke came on to Bologna that night. The other two and myself stayed over, explored Venice in the moonlight. The next morning, I walked through it's curious narrow streets for two hours before we took our luggage by gondola to the station and embarked for Bologna, which we reached safely at 6 p.m. Tuesday Feb. 11th, once more back where transportation was normal, where motor cars were abundant, where civilization seemed to be close at hand, where communication by mail seemed possible, for Serbia had only mail by special courrier service once a week, and we had been over the road and knew what time that took. In these few days, we had seen much novel and instructive and had passed through several distinct racial and political regions. My attitude of mind regarding the rights of the Jugo-Slavs and the Italians is not yet made up. In this, I am still from Missouri.

ARC Serbia Trip

February 4, 1919

You must remember there is bitter rivalry between Italy and greater Serbia, both countries claiming Trieste, Fiume and the hinterland, in order to appreciate the reasons for the difficulties of the journey.

AMERICAN RED CROSS
CROCE ROSSA AMERICANA

16 Raskova Ulica,
Belgrade, Serbia.
Feb. 4th, 1919.

I have been here exactly twenty-four hours, after a most unique trip of four days by train, steamship, and ferry boat. With my three British nurses, I left Bologna early on the morning of January 28th. It had been agreed in Rome that the rest of the party was to arrive in Bologna at seven o'clock and that we were to have joined them. The train arrived, but to my consternation, there was not a single American aboard. The A. R. C. officer in charge was to have our travel warrants with him and the necessary supply of food. The railroad officials suggested that, perhaps, they had taken wrong train, which passed through at five in the morning, and went around the upper Adriatic to Trieste. All I possessed was the four travel warrants from Rome to Bologna. I hurriedly wrote the words Venice, Trieste on these same warrants and climbed onto the Venice train. We arrived at Venice at ten o'clock, passed the officials safely, and found that there would be no boat until eight o'clock the following morning.

We all went to the Hotel Manin, spent part of the day sightseeing, and in the afternoon, while the nurses were enjoying a gondola ride, I went to the naval commander of the fort and obtained passes on the naval boat Prince Hohenlore (captured from the Austrians) and was told to be on board at half past seven. We all went to bed early and had a good night's rest. At seven in the morning, no gondoliers being awake, we formed a picturesque procession across St. Marc's square, four porters carrying our bedding rolls, the nurses and myself lugging suitcases, canteens and haversacks on our shoulders and in our hands. Fortunately, the ship was anchored near the Salute church, and we piled into a gondola which happened to be on hand and were rowed over. The new military attaché, Lt. Armstrong, and his young bride also got on board after we did, and we struck up an acquaintance. The trip across the Adriatic was very rough and unusually long, as there were still many undiscovered floating mines, and the ship had to hug the shore pretty closely. All three nurses were very sea sick, but, as I managed to keep my stomach down, I held their heads, turn and turn about, and wrapped them up in blankets as it was most bitterly cold. The Attaché lost everything but his shoes and was a most pitiful wreck.

Arriving at Trieste at three in the afternoon, we went to the Savoia Hotel, where we found the rest of the party, among whom, to my joy, was major Lee who had arrived in Rome just in time to come on. They were leaving that same afternoon, but as the three nurses were exhausted after the passage, I deemed it wiser to wait a day in Trieste and told Dr. Lee to stay with me. The A. R. C. official had totally forgotten about our passes, so we were as badly off as ever. Having gone that far safely, I trusted to luck to finish the journey with. After a good supper and long sleep, we were asked to see a naval officer who had just been operated upon for appendicities, two of the nurses of the original party had been left behind to nurse him. We also bought a good supply of food, filled two bottles with kerosene, and one bottle with denatured alcohol for our Primus cooking lamp. [We requested] the Italian authorities to reserve a compartment for the American mission to Belgrade and marched up as though we owned the world. The train left at five in the afternoon - a most dilapidated affair, with all the seats stolen by the soldiers, so that we had to sit on our bedding rolls. We five, Misses Biggs, Burke, Gerard, Lee, and myself, made ourselves as comfortable as possible, and the train started snorting up the mountain side. The sunset from Miramar across the gulf of Trieste was astonishingly beautiful, but it soon grew bitterly cold, there being no steam heat or even lights of any sort in the train. One after the other, the blankets were pulled out and piled on top of us in order to keep warm. We were shunted from one line to another until midnight, when the train arrived at a miserable station called Lovica, where ended the Italian zone of occupation and began the Jougo-Slave zone. You must remember there is bitter rivalry between Italy and greater Serbia, both countries claiming Trieste, Fiume, and the hinterland, in order to appreciate the reasons for the difficulties of the journey. As soon as the Italians realized that we were on the way to Belgrade, we were regarded almost as enemies. At Divica, the Italian terminus had been established, we were ordered out of the train and put through a stiff examination as to credentials, etc. The official looked at a pile of documents which I handed him with a flourish (none of which mentioned our destination) and passed us. We shivered in the cold waiting-room for an hour and then were granted permission to walk back to the Italian train, as it did not return to Trieste until six in the morning. I left Lee and the luggage in the station, settled the nurses with blankets, and sat shivering in my overcoat for five hours, suffering intensely from the bitter mountain air. Shortly before six, the Jugo-Slave shuttle train from Leibach (or Ljnbliana as it is now called) came puffing in, and we climbed aboard, started the precious primus lamp, and had a good hot breakfast of coffee, bread, and jam. After two or three switchings, we arrived at Ljnbliana at eleven in the morning and were welcomed with open arms by the Serbs. They requisitioned rooms for us in the leading hotel and, after a shave, lunch, and nap, we were as fit as fiddlers. Lee and I called on General Janovitch, who was very cordial and gave us his car and two soldiers to take back to the station at five in the afternoon. There we were given a sumptuous first class compartment with table, steam heat, and light. We cooked a splendid supper of hot soup, hash and potatoes, and tea and played bridge until midnight when we arrived at another terminus, Agram (or Zagreb), the capital of Croatia. This city is now a seething pot of Jugo-Slave politics, and all the hotels were full. The Serb commander finally requisitioned three rooms in three different hotels, all about a mile from the station. There were no autos or carriages to be obtained anywhere, but the soldiers finally secured a two-wheeled hand cart on which we and several Serbian officials all piled our baggage. The thing was so top heavy that it spilled over three times, my poor old suitcase finally bursting open and having to be strapped up in order to keep my clothes together. Can you picture the procession we made, all carrying a blanket over our shoulders, water canteens and haversacks slung around us, staggering along the snowy and icy road at one in the morning, the porters yelling at each other in Serbian or Croatian?

We piled the nurses in one room, which possessed two beds and a couch. I dropped off at the next hotel, and Lee went on to the third, only to find that he would have to sit on his luggage in the hall, as his room was already occupied. My room was cold, clammy, and dirty, so I didn't undress at all, merely taking off my coat and shoes. The nurses had rats hopping over their beds all night long, so we all met at eight in the morning, blear eyed, cold, and sad. I got a sled, stopped for the nurses, piled them on with the luggage, and hopped on behind, with one foot on each runner. A big bull dog didn't like my looks and made a dive for one of my legs, biting my heel. Fortunately, my heavy puttees saved me from being bitten, but I had a time maintaining my balance and kicking at him until he dropped off. On arrival at the station, we were informed that, owing to the fact that a diplomatic mission was expected on another train, ours would not leave before eleven o'clock, instead of at a quarter to nine. We were, however, comforted with the assurance that an extra coach would be placed at the disposal of the Attaché and of our mission. We were then ushered in the first class waiting room and locked in, so that other passengers might not disturb us. The nurses lighted the Primus lamp, and we had coffee, bread, and deviled ham in great style, to the great surprise and mystification of the passengers on the platform who peered in at us through all of the windows, wondering what manner of miraculous fire we were using on the waiting-room table. The people were most interesting, as refugees were flocking back to their homes as fast as the poor broken down railroads could transport them. Croctians, Gypsies, Slovenes, Serbs, Macedonian Mahomedans, Jews, all passed before our eyes, clothed in all manner of fantastic garments and carrying huge bundles of clothes and household goods on their backs, the women being conspicuously heavier laden than the men - perhaps because the latter were all too old for military service. The soldiers were all magnificent physical specimens; above the average height, broad shouldered, and deep-chested; they were real fighting men.

During the final campaign when the allied troops from Salonique broke through, at the time of the Armistice, the Serbian soldiers were eight days in advance of the French and English armies, showing their wonderful powers of endurance. The French and English officers freely admit this fact. "Give a Serb soldier a loaf of black bread and an onion or two, and he will march and fight sixty miles a day"; this statement was made to me by a French Colonel.

At about eleven o'clock, we were ushered into our coach and given two double compartments, one for our luggage and cooking, the other to sit in. The coach was a most disreputable affair, with several window panes missing, all of the doors off or half off, no lights or heat. As for the toilet, I had best not describe it's filthy condition. Fortunately, the nurses were a good old war horses without false pride or false modesty, as we had to consider ourselves as belonging to only one sex. Two soldiers were on guard, to prevent the natives from crowding in and suffocating us. The Attaché and his wife had a small, single compartment ahead of us in the same coach. Young Madame e'Attache, a born snob, did not condescend to mingle with us as she considered her Lieutenant to have a much more important mission than mine was. Her whole attitude was one of patronizing condescenstion, so we left them alone - and they never had a bit of hot food all the way up. Whenever we had to change cars, the two would rush out frantically in order to get ahead of us - only to show open mortification when I sailed majestically up with my party and obtained as good or better accommodations than they did. I used German and French all the way up and got all I asked for. Occasionally, a Serbian officer or soldier would gleefully shout to me that he had worked in Minnesota, or Gary, or in Pennsylvania coal mines, but German and French were the two valuable languages - neither of which the Attaché knew. The car back of us was filled with French officers on their way to Belgrade, so I had some interesting chats during the day. The rest of the time, we cooked meals or played bridge. This country traversed - aside from one or two mountain ranges, was as flat and ugly as Indiana or Illinois - in fact the farms, houses, and all, reminded one forcibly of American middle-west life. It was snowing or sleeping most of the time, so that one could not see much, as the windows were all coated with ice and snow. As soon as it grew dark, the huge sparks from the wood burning locomotive, gave one the illusion of a fourth of July celebration. Falling on the clear white snow, the bits of live wood and sparks made a most wonderful picture. Our guards replenished the water canteens at wayside stations, and we, in turn, fed them on hot pork and beans with a cup of hot tea. One of the soldiers insisted on my tasting some liquor which he had in his flask. It was a combination of hair restorer, eau de cologne, wood alcohol, and a dash of Vanilla - the whole combination being called "Vanilka". It burnt me clear down to my stomach. Our car went directly from Zagreb to Senslin on the river Saave, so we did not have the bother of transferring our luggage for the next thirty-three hours. At night time, the three nurses laid down in one coach, while Lee and I occupied the other one. Everybody stretched out full length, and we slept like tops until four A.M. when a perfect mob, taking advantage of the fact that our guards had fallen asleep, rushed into our coach. The nurses awoke to find a dozen brigand-like men peering at them and gesticulating and shouting in a dozen different dialects. I rushed out in my slippers, shouting "Americanski Mizion, verboten"; the guards awoke, and we got the mob into other cars. After that, we slept peacefully until 8 A.M. when the girls routed us out to cook breakfast. As the day wore on, the blizzard grew worse and worse, the rails were clogged with snow, and our poor war worn engine heaved and breathed like an asthmatic old horse. We were stuck in the snow several times, but always managed to pull through every snow drift. Lunch, tea, supper, and all of our blankets, managed to keep us fairly warm, except as to hands and feet. We played a little bridge from time to time, alternating with "pease porridge hot" in order to warm up our hands. Instead of arriving at Senlin at eleven in the morning, it was half past seven in the evening before we finally crawled into the station. There, we were informed that the last ferry boat down the Danube left at five thirty. The military post commander requisitioned three rooms in one hotel and one in another, Lee sleeping alone. We set down to our first supper in a restaurant and devoured the best roast beef and mashed potatoes I have tasted since I left Chicago. The food was excellent with delicious light Hungarian wine, much like Tokay. We all retired at nine and slept twelve hours. In the morning, we met for breakfast, piled our luggage on a sled, and walked to the ferry-boat landing. The town is semi Croctian, semi Oriental, many Mahomedans in the streets. The shoutings at the wharf reminded me of Tangiers, the same hullabeloo, the same fighting among porters, the same invocations to Allah that all of the beggars had carried our luggage. As one got 330 Corones for 100 francs, we felt like plutocrats and distributed bank notes lavishly. On the boat, I again met the British corporal and two Tommies, who come up all the way from Fiume with a month's rations for their detachment in Belgrade. Poor corporal, a real "Hoot Man" from Edinburgh, had rushed up to me in Zagreb with tears in his eyes saying "For Gawd's sake, sir, kin you talk to them bloody furriners? I cannot speak their bloody tongue, I avent ad a bloody minute's sleep in three days, and the bloody blokes won't put my bloody rations on their bloody train." I went to the commandant and told him that the rations were most important and had to go on our train - so the rations were transferred into an empty freight car, and the three Tommies locked in with their cases, to prevent them getting lost again. Corporal grinned at me and thanked me for my services; he offered me a pot of orange marmalade, and I gave him and his two men a cigar apiece - to their great delight. "Me eye, sir, we'll have somethink in front of our noses to sniff at now".

The crossing from Semlin to Belgrade, first down the Save river, and then across the Danube was most interesting. River boats, Serbian and English monitors, sail boats, and queer looking row boats, with the oarsmen in the front of the boat instead of amidships or astern, were plentiful. Our ferry boat was an old fashioned side wheeler, not unlike the Hudson river boats. The trip took about a half hour. Belgrade is a good sized city of over 100,000 inhabitants, very modern in construction, well paved, and lighted. Many of the hotels, offices, and opera house, the old Royal palace, and new palace of the prince regent are handsome buildings. From the river, the climb to the city is a steep and slippery one, but once on the plateau above, there are only slight grades. I left the party on the pier, hunted up the A. R. C. offices, and had them send for our luggage. My old chief nurse from the N.S., Miss Leete, welcomed me with joy, as did six other nurses and four doctors who had come ahead. Lt, Col., T.W. Farnum, the A. R. C. commissioner for Serbia, whom I had already met in Rome, invited me to live in the house he has rented. I am writing in my own room, nicely furnished with odds and ends of furniture left behind by the Austrians as they evacuated the city three months ago. Will probably knows about Farnum, he is an old Yale man, a banker in New Haven and one of the trustees of the New Haven Hospital.

I am not yet competent to write about the needs of the people. Of starvation, I have, so far, seen no trace; there are many beggars about, going from one A. R. C. mission to another; it is hard to get anybody to do any work, and wages are very high. Food is plentiful if you have the money to buy it with; there is manifestly a sad lack of clothing, especially among the educated classes. I visited the military hospital here, commanded by Col. Sandemeyes, a highly educated Polish gentleman from Lvoff, educated in Cracaw and who had lived in Kieff before entering the service of the Serbian Army thirty years ago. He almost embraced me when he heard my name, and I told him that my father was one of the Kieff Tarnowskys. I was invited to tea at his house, met his charming daughter - a student of law - and his older son, a lieutenant in the Serbian flying corps. A younger son is studying aviation in France. They all spoke German and French fluently, the Colonel also having a remarkable English vocabulary at his command, with pronunciation all his own. He had invited a number of Serbian notables to meet me, and I had a most enjoyable afternoon. Several of the ladies were splendid pianists, the conversation rolled along from topic to topic, a real Samover steamed on the table, and delicate ham sandwiches and cream puffs were handed around with the tea. As a special priviledge, the Colonel and his daughter whisked me out of the parlor to their private rooms and introduced me to two young Russian cousins, boys of 16 and 18 who had arrived two days ago from Moscow, after a three months' trip all alone, during which time they had constantly been on the verge of starvation. They rode or walked from Moscow to Odessa, got on board a tramp steamer which took them through the Bosphous, around Greece, and up the Adriatic to Raguse, and then begged their way across the mountains to Belgrade. The younger of the two was in bed, too weak to sit up; the older, Dimitri, was able to walk around. Both were skeletons, with sunken in blue grey wistful eyes. They only talked German, besides Russian. When I said to them "Jetzt Kormen sie gut essen, trinken und schlaffen", they both shouted "Ya wohl", and then burst out crying as though their hearts would break, from sheer joy at once more having a home with somebody to care for them and love them.

In the hospital, I saw and examined some sixty cases of Typhus fever - all among poor Serbian prisoners who had been freed by Bulgars and had to walk back to their country with little or no clothing on their backs and no food, except what they could beg or steal along their very long weary road. An English [officer] and an American Sanitary officer are stationed here, and they are doing their best to check the impending epidemic, fumigating all the clothing and killing the lice and fleas on the bodies of the poor soldiers and civilians as they come tramping in from all parts of the country.

I have sent off eight doctors to southern Serbia, via Salonique, loaned two to the military hospital here, placed another one in charge of the War Orphanage, and sent two to Sanendria and Obrenovetz on the Danube. I expect ten more doctors in a few days and will distribute them wherever they are most needed, with sufficient nurses to help out. There is no surgery, but much relief, hygiene, and infant feeding to do. The American Charge d'Affaires here is Mr. H. Percival Dodge, an old friend of Aunt Bella's, as he was attaché in Tangiers for several years, just after my mother's death. His mother-in-law lives in Nice, so they all know Michel, Ivan, Alice, and the Durandys. Mrs. Dodge is a very charming woman, doing good work along sewing-school and settlement lines. I am to lunch with them tomorrow.

Serbian trip, cont'd.

February 13th, 1919

My lunch with Mr H. Percival Dodge, the U.S. Charge d'Affaires for Serbia, again illustrated the smallness of the world. We soon discovered that we had lived in Nice at the same period of our youthful lives and had been close neighbours for several years, although Dodge was attending French school while Michel and I were at the Anglo-American college. It was most amusing to see the consternation and mortification of the young military attaché when, coming into the dining room with a message for Mr Dodge, he found me calmly smoking a cigar and drinking my after dinner coffee! I imagine hubby and wife must have had a very earnest interview that same exact evening on the danger incident to snobbery!

That same afternoon, February 6th, the three nurses, Major Lee, and myself, took the last ferry-boat across to Semlin, after bidding everybody - as I then thought - au-revoir, promissing to return in ten days. At Semlin, we obtained requisitioned rooms, ate a good supper while a good Tsigane band discoursed queer but fascinating walzes and Czardas, and then retired early in order to get enough sleep before being roused out of our beds at six in the morning. The sleigh procession on the following morning, from the hotel to the railway station, would have made a dying cow laugh! The three nurse nurses and I piled into one sled, with our water canteens and haversaca, while bully Major Lee, with all of our heavy luggage, rode in the other one. The weight of Lee and luggage was too much for the second sled, and one side broke down, so that Major Lee finished his ride balancing himself at an angle of forty-five degrees! I had to pay fifty Kroner damage to the irate driver, before he condescended to help unload. At the station, we were given two compartments, one for our luggage, the other for ourselves. The crowds of military and civilian passengers was so great however, that I had to allow the luggage compartment to be occupied; I believe we should have been mobbed if I had not consented to this. After all, we were travelling in their country, and we had no right to keep any seats empty. Of course, this cramped condition made our cooking somewhat difficult, but we managed very well and had breakfast, lunch, and supper without too much difficulty. The scenery along the Saave valley was most monotonous and uninterresting, so we played bridge most of the time, the fifth follow taking turns at reading while the other four played. At night time, our problems multiplied as we were all five in the one compartment. By geometrical adjustment of arms, bodies, and legs, with the aid of our bedding and clothing rolls which we placed between the seats, on the floor, we all managed to lie down - or almost so - the exception being Major Lee's six foot three hulk; the latter described cubist curves, everybody having some part of his anatomy in close proximity! In spite of these handicaps, we all got some sleep and, being so closely huddled together, did not suffer from the cold at all. We arrived at Zagreb on the afternoon of February 8th, and there, two routes were suggested to us, one via Fiume, the other by the old route, via Ljubliana. After much hesitation and chasing back and forth from one station to another - leaving the nurses to rest at the hotel - I finally decided to resume our journey by the old route - a most fortunate decision. We got the nurses to the new station, and then I went back for our luggage. There were no vehicles to be found at the station, and the military commander seemed rather helpless, so I rushed back to the center of the city and got a carriage. As I was superintending the loading of the luggage, there was a sudden shout and great commotion in the crowds around the station. A soldier, who turned out to be a deserter under arrest, had broken away from his guard and was making a dash for liberty. Out he shot from the crowd, with his guards hot after him. The stupid soldiers began shooting at him, and a bullet whizzed close by my head, missing me by inches! How idiotic it would have been to finish in such a way, after all I have been through in actual warfare! As far as I could see, the soldier must have made a clean escape, as he was rapidly outdistancing his pursuars when he disappeared around a corner. At seven forty in the evening, we started for Ljubliana, cooking our late supper in the compartment. At one o'clock, we arrived at the station and found that the shuttle train to Longetico left in forty-five minutes, just giving us ample time to transfer our luggage and settle down. At the Jougo-Slav-Itailan junction, we were again in luck, catching the train to Trieste an hour after our arrival at five A.M. We spent the hour making and eating a hearty breakfast. The journey over the mountains and down to the gulf of Trieste was a bitterly cold one, and we all suffered badly from frozen hands and feet. By one P.M., we were in Trieste, having beaten our outgoing journey by a whole day; we were of course very tired from lack of sleep and discomfort, but jolly glad that the trip was over. At the hotel, we met commander Esham of the U.S. submarine Chaser 324. He at once invited us to go to Venice in his boat at seven the following morning. Needless to say, we accepted with alacrity, the joy of travelling in a clean boat appealing to us most singularly. After a good night's sleep, we met at the dock and had our luggage loaded on by our smart tars. The three nurses had the time of their lives, as they were ushered into the captain's quarters, and the officers entertained them all the way across. Major Lee and I remained up on deck as the view of the Venitian coast was surpassingly beautiful. It was the most perfect day I have ever seen at this time of the year. The entrance into Venice I shall never forget as long as I live. We dined and supped at Venice, did some sightseeing in the afternoon, and then I caught the evening train to Bologna, reaching there at midnight. The following day I came on to Rome, in order to straighten out the question of the Field Hospital, which is now definately to go to Poland. This closes my Serbian campaign, as I am to devote all of my energies to the Auto-Chir, or Field Hospital.

Leaving Bologna for Marseilles

Ten Day Trip

May 24, 1919

The trip has again emphasised the extreme value of languages; I had to be interpreter, treasurer, and C.O., all in one. It was impossible to delegate a large part of the detail work to anybody else and I thanked my stabs that I could always talk the native tongue wherever I went.

Marseilles

May 24, 1919 - Marseilles

I have not written to you for two weeks because there has not been a moment to spare, and also because I wanted to wait until there was something definate to write about. The great journey is over, and I have safely accomplished what the Italians said I could never do, namely, take the heavy wheel hospital over the Appennines and Alps from Bologna to Marseilles.

My last letter was written in Paris; I arrived in Nice on May 8th and left again, in my staff car, on the 10th, sleeping that night in Genoa and arriving in Bologna on the afternoon of the 11th. Major Lee had preceeded me by two days, in order to meet the 32 soldiers who were to act as mechanicians and drivers. I found them all quartered in the Y.M.C.A. hotel on my arrival. We at once set to work to put the trucks and cars in good running order, packed our canned food, coffee, sugar, and condensed milk in the kitchen wagon and loaded up on gazoline, oil, and grease. The formal transfer and signing for the Auto-Chir took place on the 11th, after which I invited Maggiore Busi and Captain Pavanini, the Italian experts, to a final lunch. The tension was still so great that they suggested our driving out to a small village called Castellechio, in order that they might not be seen lunching with an American Officer! Even they, though quite fond of me personally, could not help harping back to Wilson and the question of Fiume, and I twice had to remind them the American officers had received formal orders not to enter into any discussions of a political or diplomatic nature. The lunch was somewhat of a strain, as you may well imagine. We took formal leave of each other, and I returned to Bologna for the final touches.

Wednesday, May 14th, we finally got off, at about eleven o'clock, as there were several false starts, the cars being new and very hard to manage; the drivers not accustomed to the heavy Italian cars. The Italian foreman smiled at me and said "you will be telegraphing for help before you reach Modena – thirty kilometers from Bologna!" We passed Modena and ate our lunch at Parma, pushing on in the afternoon to Borgo-San-Donino, 114 kilometers from Bologna. The men were tired out, but game to the core. Captain Poggioli, who had been so nice to us during the past four months, said he had some business in Genoa and asked if there was room in my car. We invited him along, and he proved to be a most enjoyable travelling companion. I really believe he came along to help us out in case we got into trouble with the people along our route! At Borgo-San-Donino, I arranged for parking of the convoy in a military park, billeted all the men in an Albergo and arranged for their supper and breakfast in the only decent hotel. Just as we were sitting down to supper, who should drive up but Maggiore Busi with three of his mechanicians. What he told me was that he had begun to worry about the ability of my men to handle the cars and that he desired me to accept the loan of the three men as far as I cared to have them go. Whether that is so, or whether it was a neat way to ascertain that we were not headed for Serbia, I leave to anybody's imagination. Of course, I accepted and invited the Maggiore to supper. He returned to Bologna that same evening. After a good night's rest, we started off again early in the morning, reaching Piacenza for lunch. In order not to injure any of the delicate instruments, the cars were only allowed to make from eight to ten miles an hour, hence the apparent slowness of our progression. My staff car ran back and forth, checking up defects, arranging for meals and billets, and, in general, acting like a mother-hen with her brood of unruly chicks! We had repairs to perform en route each day, the most serious trouble arrising from one of the solid tires of the kitchen loosening up and threatening to slip off and upset the car. We made Tortona by night; it being a British base, I repaired to H.Q. and stated my case to General Swebey, the C.O. He very kindly placed his motor park at my disposal and ordered that we be supplied with all the gazoline and oil we needed. He also sent his A.D.C. to the hotel and invited us to mess, but we had to decline on account of Captain Poggioli. Lee and I did go up to the mess rooms in the evening; we chatted, smoked, and drank until quite late. The officers were all most charmingly hospitable, a lot of bully fine fellows, all praying for an early demobilization and return to "Slighty".

May 16th, we began the Appennine climb, our first real test; we had no trouble at all and never had to tow a single car up. We lunched at the top of the Passo-di-Gioje, in a quaint trattoria, having to send the touring car back ten miles for enough bread to go around with. I had sent the motocycle on ahead to Genoa, in order to get a supply of gazoline and oil and bring it to San-Pier-d'Arena, in order to avoid doubling on our tracks; the car came to grief around a curve, both the driver and Lt Walker being thrown out and stunned. The motocycle was able to crawl in to Genoa for repairs. This mishap forced me to go to Genoa myself and delayed us quite considerably. We pushed out of the Genoa district by four P.M. and planned to spend the night in Varezze, in order to avoid any possible trouble in a large manufacturing center. We did not find food or accomodations at Varezze, so had to push on as far as Savona, which we reached at eleven o'clock, the boys dead tired and famished. They ate late supper in the station restaurant and were billeted in various hotels. The convoy had to be left out in the street as there was no other place for it, and it was too late to get military permits for the auto-park. As none of the men were in any shape to mount guard – and the Auto-Chir is valued at two million lire – Major Lee and I mounted guard all night long, one of us snatching a few groans on the driver's seat while the other paced around and around the convoy. I found walking less tiring than the seat! At seven A.M., one of the men relieved us, and we shaved, ate breakfast, and lay down for two hours, there being a broken spring on my car to repair. We made a rather late start as several of the men overslept and had to be routed up. We lunched at Noli, a quaint fishing village where the inhabitants had apparently not yet heard that Orlando and our President were no longer using the same loving cup, and reached Oneglia for the night. There I obtained permission to park the cars in the yards of a large steel plant and billeted the men in several albergos. Starting early on the 18th, we reached Vintimiglia at noon, went through the usual formalities, and then I dashed ahead to the front where I showed my papers to both Italian and French authorities and passed the whole convoy in without any trouble. The three Italian mechanics had been dropped off at Vintimiglia; they had been officious, but of no particular use. My boys were glad to see the last of them. I did smuggle my old chauffeur, Valerio, as I had promised to take him as far as Marseilles. Italy does not permit any of her subjects to enter France just now; France needs men for reconstruction work, and there are thousands of Italians out of work, eager to cross the border in order to earn good wages, but the government has adopted this means of reminding France that she needed men in order to carry on! This is not imagination; Maggiore Busi made this statement to me himself! Anyhow, I smuggled Valerio in and got him a month's permis de sejour for France, on the plea that he was necessary to me. At one P.M., we found ourselves in France, lunching at Mentone. I telephoned ahead to Nice and secured quarters for the men in one of the hotels requisitioned for furloughed men. Our worst mishap happened on the Corniche road and might have been a catastophy. Our kitchen was again the culprit, the tire suddenly slipping off, and the car tipping over on it's side, fortunately away from the sea but near a steep slide. The rest of the convoy got to Nice at six P.M., but it was midnight before the kitchen lumbered in. Part of the convoy was parked in a square, the rest in the U.S. garage.

Leaving Nice on May 22nd, we lunched at a tiny little restaurant in the Esterrel hills, using our own beef, coffee, sugar, milk, and cheese. We drove 120 kilometers that day, arriving at Le Luc for supper. As there were very few rooms in the one hotel of the village, I went to the Mayor and had him give me a list of the inhabitants who had a spare bed for the night. The town crier then went around announcing the fact that American soldiers were to be billeted with them; very few refused to accept, and then only because of illness or previous requisitions. Of course, I paid for each bed used. We also got two restaurants to cook 16 meals apeace, so that everybody ate and slept well. We all drank our own coffee and milk in the morning and made an early start, the first convoy getting off at seven, the second at eight. Passing over the Esterrel mountains, we lunched at Trets and rolled into Marseilles at 4 P.M. May 23rd, nine days after leaving Bologna. With repairs and other delays, I had estimated ten days for the trip.

We now have to prepare every car for the sea voyage, greasing and oiling all the machinery and putting vaseline on all the surgical instruments. The first boat sails on June 2nd.

The trip has again emphasised the extreme value of languages; I had to be interpreter, treasurer, and C.O., all in one. It was impossible to delegate a large part of the detail work to anybody else and I thanked my stabs that I could always talk the native tongue wherever I went.

Leaving Marseilles for Bucharest, Roumania

Curated from "Bay of Naples"

June 26 – July 15, 1919

I spoke French, German, Italian, and English in the course of the evening, thanking my stars that I had had the opportunity to learn these essential languages. I always get so much more real information out of the people; they all openly confess their inner opinions regarding our activities here and elsewhere...

June 26th, 1919 - Bay of Naples

We had a most strenuous day at Marseilles, two days ago, full of delays and unnecessary vexations. At one time I was mad enough to chuck up the whole thing and tell the Roumanians to go to h...l! In spite of the fact that the A.R.C. is presenting them with this valuable hospital, they insisted that my chauffeur and I pay full passage rates; they even wanted me to pay 2400 francs freight on my car, but this I refused to do, finally compromising on 163 francs port charges! I also managed to have the same reduction made on a Y.M.C.A. car which was going to Roumania. Besides having to pay 700 francs for my cabin, I am charged twenty-five francs a day for my food, paid in advance, and the same for my driver. Of course this does not come out of my pocket, but the principle made me mad clear through. I am becoming more and more convinced that charity does more harm than good, unless very judiciously used.

The ship finally started from Marseilles at five P.M., a heavy gale blowing. It is a small four thousand ton oil boat, very swift and correspondingly unsteady, so that my supper very promptly went overboard, along with almost every-body else's! A good nights sleep put me on my feet again, and I enjoyed the trip along the coast of Corsica very much. We passed close to the isle of Elba, made famous by Napoleon, and also saw the isle of Monte-Cristo of Alexandre Dumas fame. The Corsican coast line is very picturesque but rugged; no wonder they are a set of smugglers and brigands!

July 4th, 1919 - Costanza

On the afternoon of July 4th - which three A.R.C. nurses and I celebrated by singing the Star Spangled Banner and all the old college songs we could rake up - our good ship Imperatul Traian, arrived at the port of Costanza on the Black sea. This city is essentially a resultant of the discovery and exploitation of the Roumanian Oil fields. There are refineries tanks, pipings, etc etc, and the place has prospered accordingly. Just at present it is dead, because the huns blew up the bridge over the Danube and destroyed the pipe lines which brought the crude oil to Costanza. Most of the houses were systematically looted, even to doors, windows, and their sills; the best buildings were struck by aerial bombs at pleasure, the Roumanians having had no anti-aircraft protection during the war. The overrunning of Roumania by the German army under Mackensen, was as easy an affair as scattering an ant-hill with one's foot - the comparison is not as exaggerated as it may sound! Roumanian officers admit that they were not ready for war when they declared war - and the majority of the equipment promissed them was scattered in transit in Russia - then on the verge of revolution.

Getting Close

The heat in Roumania is stifling during the day, and the bullocks cannot travel very far. At two A.M., my eyes gave out and began to blur, so I handed the wheel over to O'Neill. He almost ran the car into a huge beam, placed across the road at the approach of a partly condemned bridge; after awakening the guard, he finally consented to remove the obstruction, but we all got out, except the driver, and walked across. The wooden bridge groaned ominously once or twice, but the car got over safely. At half past three A.M., we entered the main boulevard of Bucharest, Bulevarda Elizabet, and by four o'clock we had deposited Mlle. Diamandopol at her residence. Mr. Papavescu had piloted me to the hotel Majestic where I found one room with two beds, so I had to take the chauffeur in with me. We both slept until ten o'clock, washed up, had coffee, and then repaired to A.R.C.H.Q. There I found that her majesty the Queen had departed for Transylvania and would not be back for several days. Lt. Col. Anderson, the Red Cross commissioner to the Balkans was also out of town. I consequently made plans to start for Costanza, in order to meet my convoy. My real difficulties now begin; I am afraid that many of the wooden bridges will not bear the weight of the heavy camions!

July 9th - Danube

After a short interview with Col. Anderson, who had returned the night before, I started out at ten o'clock for a four hundred kilometer ride to Costanza. We arrived at the Danube at noon, at a small town called Olternitza. There was no ferry-boat service, as yet, and I had to chase all over town to find the port commandant, without whose authority nobody would lift a finger. Lt. Andrei turned out to be a man of energy and resourse. He hurriedly finished his lunch and drove back to the port where he had three ordinary row-boats lashed together, side by side; across these he placed two heavy planks, at the proper interval, so that the automobile wheels could run on them, and had my impromptu raft ready in less than an hour. The chauffeur and I lunched in the meantime, using our Primus lamp and food supplies. Amid shouts of warning, the car slid onto the raft, and, with seven oarsmen paddling away like mad, we slowly but surely made our way across, the current carrying us half a mile down-stream. There was a good sandy beach, fortunately, so that with the help of two heavy boards which the boatmen had brought along with them, we managed to run the car onto solid sand and up the bank to a small path which led to the former Bulgar village of Turtucaia, where we had to report to the military commandant before proceeding. The boatmen charged about twenty dollars, a very reasonable sum, considering the fact that they had to get back to Oltenitza and undo their raft. The military commandants all refused graft--because I could talk to them either in French or German. I had been told that the crossing of the Danube would cost a hundred and fifty Lei in graft, plus the boatmen's charges. I am becoming more and more convinced of the fact that people are cheated because, not knowing a word about languages, other than English, they tempt the natives beyond the power of resistance. At any rate, there we stood, gazing at the Danube, after our most unique experience!

July 10th - Japanese Steamer and the Floating Crane

Late that same afternoon, July 10th, the S.S. "Fukui-Maru" hove in sight, and I was delighted to see my entire hospital on board. My five drivers had had a very comfortable trip, sleeping in the ambulances and cooking all of their meals in the rolling kitchen. The steamer is a Japanese one, loaned to the Allies for three years, and transporting army supplies for the French army of the Orient. My entire convoy of eighteen trucks was carried to Costanza free.

To my consternation, I now found that the steamer cranes were not powerful enough to unload the trucks; they had used a heavy floating crane at Marseilles for the loading process! The floating crane at Costanza had been sunk by the Germans and still lay in the bottom of the port. The French commandant wirelessed the French admiral at Constantinople for permission to send the ship up the Danube to Galatz; it took two days to get an answer; in the meantime, some three thousand tons of French freight, had to be unloaded at Costanza in order to make the ship light enough to pass up the Sulina channel of the Danube, which only has a depth of twenty-one feet. I decided I might as well load my car on board and go with the convoy. We sailed July 13th in the afternoon. I was given a very nice cabin and ate my meals with my men, the cook being a theological student at Northwestern, named Odell. We anchored at Sulina the next morning, but [we] were delayed all morning in order to build an addition to the rudder, the ship not being built for quick turns. Measurements of the ship also had to be taken, in order to estimate the channel charges. The Danube navigation is an international affair, all vessels having to pay fixed charges for passing through the channel. We could not make Galatz before dark, so had to anchor in mid-stream some twenty miles away. The night was a beautiful one, the reflection of the full moon on the silent waters, the flat marshes all round one, with the distant Dobrodgea hills and the fireworks the French were sending off at Galatz, in celebration of the 14th; the mysterious Jap sailors, off-duty, with their kimonos on while their clothes were hanging on ropes and stays, drying in the breeze, made a pot-pourri picture as fascinating as it was unusual!. It all seemed like a delirium tremens of a dream; too uncanny to be actually true!

July 15th – Galatz and the Coal Crane

By seven o'clock, July 15th, we had anchored in mid-stream opposite Galatz. The pilot told me that there was no dock available at the time, and that the ship would have to remain anchored for a day or two! I jumped into a row-boat, called on the Roumanian commandant, and got permission to have the ship tie up to the military dock, in order that I might unload my car and proceed to Bucharest to report. Up went the anchors again, the Jap captain looking at me with much new respect, almost awe! I got my car ashore, drove to the port, and found that there was no crane strong enough to unload the camions, so arranged to have the steamer proceed still further up the river to Braila, where I was assured that there was a heavy coal crane which could lift ten tons at a time! Also arranged to have eighteen flat Ry cars put at my disposal at Braila, to transport the hospital to Buchapest. By that time, it was noon, so my chauffeur and I lunched on board, ran the car onto the Ferryboat, and started off for Braila, to duplicate the trip we had taken the previous week. The journey was an uneventful one, and, there being only two of us, we made better time and arrived at Bucharest at half past eleven P.M., instead of four A.M. Another bridge had given way in the past week, so that we had to make another detour; the heavy trucks could never pass over them without crashing through.

I have gone into details, in order to show you how impossible it is to hurry matters, and how many unforeseen problems I have had to meet. My troubles seem to be over, now, unless the freight cars fail me utterly.

July 16, 1919 – The Queen

July 16, 1919

As the hospital is a special gift to the queen, she had me show her the set of photos taken in Bologna, and explain the technical parts. I am to teach her court physician, col. Mamulea, the technical parts of the Auto-Chir. As I was leaving, the queen invited me to a musicale, to be given that same evening...

July 16th, 1919 – The Queen

I had the pleasure of being invited to tea, to meet H.M. Queen Marie. She received me most graciously and, while never forgetting that she was queen, she had the happy faculty of making you feel quite at home and almost on informal terms. Physically, she is a stunningly beautiful woman, with a very strong face, full of determination and energy, a happy blending of the English and Russian dynasties. As the hospital is a special gift to the queen, she had me show her the set of photos taken in Bologna and explain the technical parts. I am to teach her court physician, Col. Mamulea, the technical parts of the Auto-Chir. As I was leaving, the queen invited me to a musicale, to be given that same evening at the home of her principal dame de la cour Mme Procopui. It began at ten o'clock and lasted until the queen departed at one A.M. The entire diplomatic corps was there, so I met some very interresting people, amongst others the new Polish minister, who immediately tried to make me out as a distant cousin of his. I was also informed that the Roumanian army had a colonel Tarnowsky, who had fought so bravely that, when finally captured by the Boches, he had been allowed to keep his sword, as a special appreciation of his bravery. The musicale was a very good one, the best singer being a Russian refugee, the violonist a fascinating Tzigane who marched right up to you and played into your very soul; the ladies seemed very much épris with him! The queen sat in the only corner of the drawing room having no door or window behind her; mere accident - or fear of a bullet or dagger from behind?; everybody was closely scrutinised as they entered the house; these are troublous days for royal heads! As each guest arrived, he walked to the Queen's corner, bowed low and kissed her hand; if she spoke, the guest would answer, if not, he backed away and mingled with the other guests. I spoke French, German, Italian, and English in the course of the evening, thanking my stars that I had had the opportunity to learn these essential languages. I always get so much more real information out of the people; they all openly confess their inner opinions regarding our activities here and elsewhere, and I have a full realisation of the crude bungling mistakes our people have made. We do have a vulgar love of display; we do love to advertise our gifts, to talk in terms of dollars, to make invidious comparisons, which rudely shock the sensitive nerves of the educated classes in Europe. It is a great pity, but we are very young and unsophisticated.

July 18th – Roumanian Visit

Met Col. Mamulea at the Royal Palace and talked about the emplacement of the Auto-Chir with him. The palace rooms are very Louis XIV in style; nothing particular about them; a few good pictures by their best artist, Grigoresco, representing typical Roumanian scenes, and scores of priceless Persian and Turkish carpets and rugs. In the evening I dined with Col. Anderson, the A.R.C. commissioner to the Balkans, to meet General and Mrs Prezan. The general is chief of staff of the Roumanian army, a most charming and interresting man, speaking French like a native. Mrs Prezan struck me as the best type of patriotic woman I have met here, so far. She was fully alive to the hideous contrast between the blatant luxury of Bucharest and her poorly clothed and worse fed soldiers at the front; between the feudal Boyard and the ragged, ignorant peasantry. I was very much impressed with her and hoped that she was not alone in her just appreciation of facts. Much cheered up by a telegram from Galatz, announcing that my convoy was loaded on freight cars and had started for Bucharest, I spent the morning visiting the War exposition which was formally opened that day. I was taken around by two Roumanian officers, who explained the battle fields to me and showed me the principal points of resistance and the site where the Germans finally got through and forced a general retreat to Iassi. They attribute all of their misfortunes to the Russian revolution - which is undoubtedly correct. Nevertheless, they generously admit that, as long as the Russian troops held, Russian aid had been unstinted in quality and quantity. The retreat, and the winter following it, cost Roumania almost ten per cent of her population, starvation and typhus causing the greatest losses. Nowhere along the battle line is there any evidence of intense fighting; which is easy to understand when one remembers the marked difference in organisation and especially in artillery between the army of Mackensen and that of the Russo-Roumanians. Bucharest is not a bit damaged; it was simply evacuated. The Roumanian soldier is a docile, polite, simple soul; he has fought bravely and with as much intelligence as he was able to receive from his officers. When I see the Roumanian officers strutting along the Calea Vittoria - the leading to-see-and-be-seen street of the capital, corseted, rouged, powdered, and hair-oiled, it makes me ponder very profoundly. I assure you that I am not exaggerating their make-up! The society women all make up like shop-window mannikins; they are often very beautiful, especially those with mixed Greek or Gypsy or Slav blood. They are essentially feminine to their finger-tips; I have met no sexless types, so far, and doubt whether there are any genuine suffragettes - although at a musical review, which I saw the other evening, they had a good take-off of a suffragette meeting, carrying banners on the stage which read "down with the pants" and "up with the skirts" - both of which statements were variously interpreted by the hilarious audience! At this review, most of the songs were political ones, either for or against various members of the government; the whole city seethes with politics.

Sunday, July 20th - Royal Lunch

An invitation came to me at the hotel, to lunch with the King and Queen at the country residence of one of the old Boyard families, the Otetelecheanus. We first drove to the country palace, some five miles out of town, a huge pile of buildings originally a monastery to which various wings had been added. We were ushered upstairs by court servants in their gorgeous black-and-gold livery, and presented to Mme Olevaru, the dame-de-cour of that day. She informed us that the King was out, attending a special meeting, but that we would start as soon as he arrived. Pretty soon, the guard of honor came out in the courtyard of the palace, and the bugles blew the royal salute as the royal car drove in. In a few minutes more, we were told to get into our car, and to follow the King as soon as the bugles sounded again. I craned my neck and saw the king and queen get in their car, the King driving himself, with the royal chauffeur at his side, the Queen and Mme Olevaru behind. We, Col. Anderson, Mme Sevescu, president of the Roumanian Red Cross, and I, followed along for twenty miles. As we approached the feudal village of Otetelecheanu, the road and bridges were festooned with flowers in honor of the royal visit, and the entire village was out to shout a hurrah for the royal carriage and the American one as well. On arrival at the house, I was presented to the King; a very shy modest, but earnest and genuine man, evidently extremely nervous and ill at ease with his people. He talked freely on banal subjects, laughed genuinely when the irrepressible hostess introduced him to her latest pet dog, whom she had named "Sale Boche", and did not seem to mind any disparaging allusions to the Teutons, in spite of the fact that he is a Hohenzollern- Sigmaringen. After a visit around the house - a huge rambling affair without any particular style of architecture, the inevitable Zacouska were passed around, the chief beverage being Zuica, the national liqueur, distilled from prunes. The dinner was an orgy of meats, vegetables, sweets, and fruits - all from the estate. Mme O. kept apologising for the absence of certain dishes, bemoaning the fact that she only had "vingt-quatre domestiques, tous plus bêtes les uns que les autres"! What banquets must have been "dans le bon vieux temps, quant on pouvait vivre" - I am quoting the hostess - I leave to the imagination! When through, I wanted to lie down on the wooden terrace, just off the dining room, but had to smoke cigars and bite my lips to keep awake, chatting with the King or his A.D.C. At three o'clock, the villagers all arrived with fiddle and zither at the head of the procession and danced their national dances for our benefit. The music was weird tzigane, the dances, round dances, although every once in a while, a couple would enter the ring and twirl madly around for a few minutes. Men and women had a flower stuck back of their right or left ear; the men dressed in white embroidered trowsers with a long white overshirt hanging down like a nightgown to the knee and a black or white fur cap - the stylish ones a felt hat which spoilt the picture. The girls - the married women have to give up dancing as they all have babies at the breast or en-route - have wonderfully embroidered waists, with rows of gold pieces strung around their necks, gay petticoats and scarfs of lace or silk over their hair. The faces were not handsome, but they looked healthy and honest. The men are much finer looking than the women. All of the children were allowed on the terrace, the Queen giving each one a piece of candy. There was also a distribution of cigarettes among the men. The best looking girls were also allowed to come up and kiss the Queen's hand, and two young men made the usual halting speech of fidelity and devotion to the king. At four o'clock, we had to repair again to the dining room where tea, wines, cakes of all sorts, in quantities sufficient to stock a cake-shop, and jellies were served. Mme O. offered a huge sweet cake to Col Anderson - at the same time asking him for another enormous quantity of sugar "por mes pauvres"! There were some twenty dogs in the dining room, and the amount of cake they consumed made it very apparent that, unless the generous A.R.C. kept giving more sacks of sugar, "les pauvres" would indeed suffer cruelly! Among the other guests present were the two Mlles Robesco, whose father has been grand master of ceremonies for over thirty years. They were both very charming, highly educated, and most refined; one a raving beauty, the other cosi, cosi. It was a real treat to converse with them.

We started back for Bucharest at five; the Queen invited me to attend the first unreeling of a series of films taken during the recent royal visit to the newly acquired lands in Transylvania. They were to be shown in the grand ball-room of the city Palace. Nine o'clock was the hour. The entrée, removal of cap and gloves, and marching upstairs took a half hour, at the end of which the master of ceremonies tapped the floor with his sword and had the guests range themselves on either side of a long hall, which led to the ball-room. The King and Queen, followed by the dames-du-palais and gentilhommes de la cour, all in grande tenue, entered and bowed, right and left, before taking their seats. The films were at once shown, excellent ones, especially of the Queen who has evidently practiced the art of "registering" her features on the film for the benefit of her people, who all adore her. The processions of Transylvanian peasants were most interresting, and the loyalty of the people, judging from the waving and bowing, must have been most pleasant to their majesties. Something happened to the machine after half of the films had been thrown on the screen, so the Queen ordered the chairs removed and ended the evening with a ball. I danced with the Mlles Robesco, who bemoaned the fact that other Americans had totally forgotten how to waltz. It really was a treat to have polkas, mazurkas, and waltzes taking the place of our hideous cake-walk dances - although the Roumanians occasionally tried some of the new-fangled steps. At one o'clock, their majesties rose and bowed right and left as they passed out of the ball-room; the music stopped and everybody went home. It really was a most interresting evening; I must say I enjoy a cosmopolitan crowd! I have been offered the post of Commissioner to Roumania, but declined with thanks as I am already one month late and am anxious to be up and on my way home. From a purely social point of view, the offer is a tempting one, but I must keep up my surgery, and my fingers are stiff from disuse.

July 26, 1919 – Bucharest

July 26, 1919

By the afternoon of the 24th, the hospital was finally placed in the palace grounds, the keys handed over to the Roumanian doctor and I heaved a sigh of relief to think that the responsibility of two milion francs of material was off my shoulders!

July 26th, 1919 - Bucharest

This is my last day in the capital, as I hope to start out in the morning for my long drive to Belgrade, where I have to demonstrate and deliver a small field operating room and X ray apparatus to the Serbian army, before returning to Paris.

Monday, July 21st was spent with the court Physician, Col. Mammulea, going over the different parts of the Auto-Chir and arranging for it's temporary storing in the Royal Palace grounds of Cotroceni, five miles from the city, where their Majesties always sleep. In the evening, I was allowed the priviledge of reading the manuscript of an unpublished book of the Queens, called the "Voice on the mountain" - A story for those who understand. It is a very fascinating mystic fairy tale, written during the darkest hours of the war. Slavic and fatalist in it's sad tones, it revealed a most sensitive, poetic soul. One statement particularly impressed me: "It is only the things that float lightly over the earth that are joyful; the moment one digs a little deeper, one finds nothing but tears". I finished the manuscript in one night!

The following day, July 22nd, the Auto-Chir arrived on a freight train from Galatz, the roads not being passable on account of the broken or weakened bridges. I lunched with the Sevescus, a very select affair with two of the dames de la cour and two French Colonels. In the afternoon, I invited Mlles Robesco for a drive through the interesting parts of Bucharest. They took me all over the Carol Ist park, where in 1908, the 40th anniversary of the late king's reign was celebrated. It is now used for special festivities and for Olympian games, there being an excellent stadium. In the Robesco home, one of the best in Bucharest from an architectural point of view, I was shown a complete collection of autigraphed photographs of all the members of the Hapsburg and Hohenzollern families, now carefully hidden away at the King's request. I was also loaned a copy of the Queen's book "Lily of Life", a quaint fairy love-story. Again I read almost all night long, in order to finish the book and return it.

By the afternoon of the 24th, the hospital was finally placed in the palace grounds, the keys handed over to the Roumanian doctor, and I heaved a sigh of relief to think that the responsibility of two milion francs of material was off my shoulders! Their majesties invited me in the evening to attend a special benefit performance of moving pictures and native dances given in one of the open-air theaters. I sat immediately in front of the royal box and enjoyed the evening immensely. The songs and dances were particularly charming; the enthousiasm of the people for their sovereigns very genuine.

The following day, a messenger came to the hotel with a request that I lunch at the palace with their majesties. I barely had time to put on my best war-paint; booted and spurred, I jumped in the car and was whisked away. After being ushered into one of the drawing rooms, I was introduced to the dame de la cour of the day, and to the king's A.D.C., Col Vivarescu who whispered in my ear that my seat was to be at the Queen's immediate right! I immediately offered up a silent prayer of thanks to the Mlles Robesco for having enabled me to read two of her Majesties books! At half past one, the King and Queen entered, shook hands with me and bowed to the other guests, and we entered the dining room after them.

There were twelve people at the round table, the King and Queen sitting opposite each other, the retinue ranged to right and left, a Roumanian general at the Queen's left, and two poor little lieutenants, half dead with fright, shivering at either side of the King! They ate hardly a thing and never opened their mouths, being whisked away by the A.D.C. as soon as the meal was over!

The Queen was most charming and gracious, very soon giving me an opportunity to discuss her books; she is eager to have Loie Fuller play the "Lily of Life", so we discussed scenes and color schemes and lights and shadows until I almost imagined myself an impressario! The king entered quite frequently into the conversation, sometimes in English, more often in French, and occasionally in German, the courtiers grinning approval around the table. It was really a very interresting luncheon, and will be one of my pleasantest souvenirs. After lunch, we all repaired to one of the drawing rooms where the King and Queen talked with me for half an hour before retitring to their private appartments. Both expressed regret that I could not remain in Roumania longer. I have again been urged to reconsider my decision to depart; all sorts of flattering offers being made, but it is high time that I resume my normal life again; I really have been terribly homesick in Roumania, in spite of - or perhaps because of the exceeding hospitality of the best classes here.

No to a Queen

Were these the good old times of the middle ages, I would be in a dungeon awaiting my turn with the hangman; I have said No to a Queen! I was called to the palace this morning, and the Queen in person urged, almost begged me to stay and go to the front as commanding officer and surgeon of the Auto-Chir. She argued that, as I knew all about the hospital and could speak several languages, I would be the best person to take hold of the unit. She is so full of personal charm and magnetism that it was the hardest thing in the world to stick to my original decision, but my orders stood me in good stead and I pleaded military necessity. Colonel Anderson had, just previous to the final interview, assured me that he would telegraph H.Q. in Paris for military permission to remain here, but I assured him that I could not remain any longer. He then got into communication with the Queen and made it a personnal matter, which was very mean of him.

July 27, 1919 – Bucharess

July 27, 1919

...the two compared notes and finally allowed me to proceed. I then got peeved and told them that they should give me an order in Serbian, enabling me to pass through their lines. I explained that I was bringing an operating room and X Ray apparatus, a gift of the American people to Serbia, and that, if they could not read my orders, it was no fault of mine!

July 27th, 1919 – Leaving Bucharest for Belgrade

Having sent off the heavy camion, containing the X Ray apparatus and operating room early in the morning, with orders to keep on going until I caught up with it, I remained over at the special request of Col. Anderson, as he was giving a garden party, with a series of moving pictures taken for publicity in America, to which the King and Queen had promissed to come. It rained off and on all day long, but cleared up sufficiently towards evening in order not to spoil the affair. A large gathering of Roumanian officers arrived by nine o'clock and gave the Stirbei palace gardens a very festive air. A gypsy orchestra played under the trees, and a good buffet catered to the inner man. Their Majesties arrived at ten o'clock, and the pictures were immediately thrown onto the screen. Both the Queen and Col. Anderson were "registered" quite a number of times; the rest of the pictures depicted the visit of their Majesties through the much disputed Banat region, which I was to traverse a couple of days later. At midnight there was dancing, while the King and Queen partook for refreshments. At one o'clock they departed, the Queen shaking hands with me and wishing me bon voyage. She also promissed to send me her autigraphed photograph.

Driving to Belgrade I elected to drive the staff car myself, with my chauffeur and the camion driver in the heavy car. It was necessary to have a camion, in order to carry sufficient oil and gazoline for the 560 K.M. drive. The country traversed was a wonderfully fertile one, the farms quite American in size, though devoid of farm-houses; the winnowing of the grain being done by excellent steam-driven machinery. The villages, though plentiful, were all primitive in the extreme, chickens, geese, pigs, women, and children all living promiscuously in adobe or thatched houses, one story high. The number of totally naked children running around was most noticeable; the older children frequently only had a canvas sack on their backs, with holes for head and arms! None of the men and women in the fields, or in attendance on the winnowing machines, wore shoes; the women however all had beautifully embroidered blouses and petticoats. In some districts, the petticoat was, that is to say, had been white, with streamers of gay colored tassels hanging down all around, giving them a very fancy dress-ball effect, which was pleasing to the eye. The handsomest types are invariably found among the gypsies; girls and boys both have glorious soulful black eyes, classical Grecian profiles, jet black hair, and dark brown complexions. They all have shifty eyes and nimble fingers, and it was always wise to keep all of one's luggage in direct vision whenever the car halted! Both sexes begged for tobacco, the women enjoying cigarettes and pipes as well as the men. After driving 116 K.M., we arrived at the town of Pitesti, where we lunched, cooking our own coffee and eating potted beef, jam, and bread. By five in the afternoon, we caught up with the operating-room camion, the two drivers having spent the night in the car itself, rather than taking any chance of having things stolen from it if they had accepted the hospitality of any a of the natives. We drove together for several miles, and then I went on ahead to locate a detour of fifteen K.M. which we had to make on account of a blown-up bridge. At the little village of Chanchareni, while I was trying different languages on the natives, in order not to miss the turnings, a dear old man came up and asked me if I could "parler Francais". Upon my very affirmative answer, he begged me to honor him by having supper and spending the night with him. As it was then almost eight o'clock, I accepted with alacrity, especially as he could also bunk my two drivers. The last camion lumbered up an hour later, but the two drivers again elected to drive further on and sleep in the country. While two drivers prepared and ate their meal in the kitchen, I was introduced to the wife, a dear old white haired lady. Monsieur Nicholas Pavet, a native of Lyon, had come to Roumania 40 years ago, with his wife, a Belgian from Liège, in order to superintend the exploitation of vast crown forests which were then to be cut for the government. As part payment, he received a huge tract of crown land near the little village where he was now living and became a gentleman farmer. He had retired some fifteen years ago, and was living quietly with his three sons, all of whom are officers in the Roumanian army when the war broke out. A few months later, the Germans, under Mackensen, overran the country, and he found his house filled with staff officers, he having been denounced as the wealthiest man in the village with an exceptional wine cellar! He took great pride in showing me the cunningly hidden trap-door, between the ceiling and rafters, where almost his entire wine cellar reposed in security, with all of his wife's jewelry and his cash and stocks! Both said that the German officers behaved very well, seemingly much impressed with the old couple's repeated statements that they were "tres fiers d'etre Francais et Belge". All of his horses, carriages, cattle, tools, etc, were requisitioned, and are lost for ever.

Banat

The supper, though cooked by Roumanian servants, was most excellent; as for the Zacouska, wines, and liqueurs, they were all old and rare vintage. We ate in the garden, in a lovely bower, with vines and fruit trees all around us, and chatted until late at night; both wanted to hear all about the armistice fêtes in Paris, and all about the court news at Bucharest. I was given an excellent, spotlessly clean room and bed to sleep in, and went to sleep as soon as my head touched the pillow. I had driven 290 K.M. that day, and my arms were pretty tired. In the morning, I had breakfast with my hosts, and then Madame brought me a lovely old hand embroidered bureau cover "pour Madame votre femme"! When I asked them if there was nothing I could do for them, they timidly inquired if I could spare a little gazoline! I gave them three gallons, which seemed to fill their hearts with joy. We exchanged cards, shook hands all round, and I drove off feeling that I had met a wonderfully good couple of mortals. During the first ten K.M., beyond Chanchareni, I counted three bridges which had partially crashed through during the passage of the operating camion! I was therefore not surprised to meet the car within an hour. Twice they had been hauled out by teams of oxen, the natives taking American cigarettes in full payment of their trouble! As they now had had two days and nights without any mishap, I decided that they could shift for themselves, as they had their orders with them and an ample supply of gazoline, so drove right on through Turnu-Severin to the present zone of occupation of the Roumanian army in the Banat. Here I was challenged by a Roumanian soldier, who passed me without looking at my papers. The Peace conference had decided, just two days previously, that the Roumanians were to occupy Temesvar in the northern Banat, while the Serbians were given a new slice of southern Banat. Both sides are dissatisfied, both want the whole Banat, and both feel that poor France is personally responsible for the change - for no other reason than that the French army of the Orient has been given the most ungrateful task of sitting between the various warring factions in the Balkans, in order to prevent a fresh war from starting! King Solomon himself could not decide which part of the Banat, or of Macedonia, or of Bessarabia, or of Transsylvania should belong to Russia, Austria, Serbia, Roumania, or Bulgaria! In one of the villages where I had to stop, on account of puzzling cross-roads, a dear old Orthodox priest came running out of his presbytery and asked me if I would take him to his other parish, six K.M. further on. I took him with pleasure, as it gave me an opportunity to get his views. He was overwhelmed when I told him that I was an American; he had never seen one before in his life, and he was "so proud to shake the hand of one of those wonderful people who had come to blood-stained Europe in order to stop the war! I asked him in what language he preached, "In one, I preach in Serbian; in the other one, I have to use the Roumanian language!

How does it happen that you talk German so fluently?The majority of the people are Swabian, talking a German dialect!" And there ye are, as Mister Dooley would have said! At another place, I chatted with a Roumanian peasant, who had been in the Austrian army and could talk German. He said "I am a Roumanian by birth, yes, but I hope that Serbia keeps the Banat. My countrymen were here until three days ago; they took everything away, cattle, chickens, pigs, and grain without paying for anything; we don't want them back again!" It is all a mass of racial, political, and religious lies; nobody tells you more than half a truth - except the peasants when they can talk without danger of being overheard! All they want is to have good crops, to sell the surplus at a good profit, and not to have to pay graft. What difference can it make for them, whether they are governed by a Pasha or Vizir or Boy - who had bought the appointment from the Sultan, intending to repay himself many times over by squeezing the peasants - or by a Serbian or Roumanian prefect who carries out the time-honored custom of demanding as much as he can get out of the people, pocketing the lions share for himself!

Blue Danube

From Orsova, the Danube valley narrows down to a huge slit, called the Iron Gates of the Danube. It is a drive of 140 K.M., the road hanging over the river-bed, hewn out of solid rock, and just as wildly picturesque as the Corniche road. It is so narrow however, that one has to drive slowly and toot one's horn almost constantly. At various strategic places, are the ruins of the old forts, built by the Austrians or Serbians, in order to check the Turks. The river itself is very sluggish, and looks more like a huge elongated lake than a river. As to it's color, the man who wrote about the "Blue Danube", must have had an advanced case of delirium tremens when he coined the phrase! I have seen the mouth of the Danube, steamed up it for three hundred miles, crossed it at Oltenitza, at Pancshova, and at Belgrade, and walked along it's banks at Vienna in 1913; never have I seen it any other color than a dirty grey! At Bazias, a much war-battered village where there are valuable iron ores, I came upon a bataillon of French poilus, keeping the Serbians and Roumanians apart! The men were all very much fed-up with this unwelcome police duty, all the more as they were hated by both factions and had to keep strictly to themselves and not mingle with the natives. When I told them that I was on my way back to Paris, they all shouted: "Bon-Dieu de Bon-Dieu, mon Colonel, prenez nous avec vous!" They certainly were the most homesick lot of Frenchmen I have met anywhere! Can you imagine what it must be like, to live in a destroyed village, with high cliffs on one side, the huge, slow, sad river at your feet, nothing to do but the perfunctory daily drill, and suspicious natives hanging around you all day long? It is enough to test an angel's patience! A few miles further, at the village of Langenfeld, a Serbian soldier came running out of the guard house and challanged me. Pretty soon an officer appeared who spoke Italian. He looked at my papers, complained that they were written in English, had me translate the gist of the orders for him, and became wildly excited when I told him that I had come from Bucharest! "Where is the Roumanian army now; is it true that there is a revolution in Bucharest and that seven thousand Roumanian soldiers have been killed by the Hungarian Bolchevicks in Transsylvania?" I told him I knew nothing about these matters, and again explained what my peaceful mission was about. He finally told me to go ahead, but said he would telephone to the next village and notify them of my coming. He seemed to think that I was trying to go north to Temesvar, which the Serbian army had just been forced to evacuate. Surely enough, no sooner had I entered the village of Weiss-kirchen than two Serbian soldiers, armed to the teeth, stood in the middle of the road, blew a whistle and signaled me to stop. As my car did not come to an immediate halt, the whistle was again angrily blown, and one of the soldiers slipped a cartridge in his gun and pointed the latter at me, while I frantically applied both breaks and came to a halt! In broken German, he told me to go at a walking pace, jumping on the running-board and keeping his bayonet within tickling distance of my stomach; his companion did likewise with my two chauffeurs in the camion, and all of the village people came running out to see a good killing! At the other end of the village, I was told to stop my car, and ushered into the Commanding officer's quarters while my drivers were guarded outside! Again, the same line of questions were put to me; again, the officer - who spoke German fluently - complained that my orders were not written in Serbian. He called in another officer who spoke French; the two compared notes and finally allowed me to proceed. I then got peeved and told them that they should give me an order in Serbian, enabling me to pass through their lines. I explained that I was bringing an operating room and X Ray apparatus, a gift of the American people to Serbia, and that, if they could not read my orders, it was no fault of mine! They calmed down and apologised for the delay but said that I would have to drive on to H.Q. at Versetch, twenty K.M. further on, for the "laisser passer", when I got out into the street, my drivers almost wept with joy; they had actually been worrying about my life! Off we drove, the villagers showing evident disappointment at the banal dénouement of the incident! At the next village, we were again halted, but, instead of being examined, a Serbian gendarme said that he was going on to Versetch that night, and asked if he could ride with me! As his bayonet looked nice and sharp, I invited him with alacrity, and he hopped in beside me, his bayonet still in his gun, ready for execution! It was then after nine o'clock in the evening; the excitement of the past two hours had made us forget all about our supper! As soon as we came to a well, we stopped, made our coffee, and ate corned beef and bread, the guard watching us curiously. We gave him a cup of coffee and a jam sandwich - which seemed to make him more humane. We were challanged once more, but this time our guard had the pass-word and we were not halted.

Vienneese Hospital

We arrived at Versetch, a town of thirty thousand inhabitants; the guard directed us to the leading hotel and bade us a polite good-night. The hotel prooved to be full up; the proprietor telephoned around to other hotels and found them all filled up; as a last resort, he called up the police station where they told him that I would have to report there in person before they could assign me to any requisitioned rooms. I was just leaving the hotel, when a very nice young gentleman came up to me, and, in broken English, invited me to spend the night at his house. He also added that he could put two extra cots up for my men. I accepted with pleasure, as it was then eleven o'clock and we were all pretty tired out. We first drove to Mr Bauernfreund's house, where I put all my luggage then repaired to the police station where we placed the two cars in the inner courtyard for safe keeping. My host introduced me to his wife, a most charming Vienneese woman; we had a liqueur and a smoke, and then retired for the night. In spite of the fact that the one baby of the family was teething and crying part of the night, I slept like a log. At breakfast table the next morning, Mr Bauernfreund unburdened himself of some of his anxieties and worries. He, also, is a Vienneese, highly educated and intelligent. He said: We Austrians look to America to repair the great injustice which has been done us as the result of the war. Versetch is a town of thirty thousand inhabitants, of which five thousand are Serbians, two thousand Roumani, ten thousand Germans, and the rest Magyars, Jews, etc. The German names of the streets were all changed to corresponding Serbian ones, the shopkeepers were busy effacing their German signs and putting up painted linen signs, written in Serbian, and the whole city was trying hard to look like what it was not meant to be! The farmers we met on the excellent 67 K.M. long road between Versetch and Pancsova all spoke Swabian German; at Karlsdorff, a prosperous looking farmer and his wife begged me for a lift to the river as there were no trains running. They spoke excellent German. Pancsova, situated at the confluence of the Temes and Danube rivers, seemed a prosperous town of about fifty thousand inhabitants. Everybody I spoke to answered me in German; the peasants and small tradespeople were a mixture of Roumanian, Serb, Greek, and Jewish races. The

French were again doing police duty there; I had a long chat with the French navigation commandant; he and his men were praying to be sent back home to France; the Serbian papers of Belgrade had recently published an editorial in which it was stated that the Serbs were tired of seeing foreign soldiers - Peace conference or no peace conference; the French staff got peeved and packed up the same day for Temesvar. Officially, an apology was sent to the French, but the truth persisted nevertheless. The plain fact is that the Balkans, true to their secular tradition, are tired of eight months' peace and would undoubtedly be at each other's throats by this time, were it not for the aforementioned policing - to which they all object most strenuously. A British constable, trying to prevent a fight between Limerick and Cork Irishmen, would be as popular as the French soldiers in the Balkans now are!

We lunched at Pancsova, after placing my staff car and the camion on a barge which was towed by the passenger ferry-boat. The trio up the river took a little over two hours; evidences of the intense Austrian bombardment were evident in all of the villages and towns we passed. Belgrade itself has improved wonderfully since I was here in February. The city is filled with people; most of the bombarded houses have been repaired, and there is a general air of normal prosperity which was most encouraging to behold. My good old friend Col. Sandemeyer is no longer chief surgeon of the military hospital; most of the members of my old unit have departed for the States, and new members have arrived to take their places. The government is having some difficulties; there are many rumours of a parliamentary crisis; it seems that the new sections of greater Serbia are slowly but surely gaining the ascendancy; the Croats, especially, are proving themselves the masters of the less experienced old Serbs!

It was a great pleasure to see the U.S. Minister Mr Dodge again; he was feeling quite cheerful over the present political outlook. I am to be presented to the Crown Prince tomorrow, if possible.

August 1, 1919 – Belgrade

August 1, 1919

...my car. It had originally belonged to the Auto-Chir, but my orders from the A.R.C. in Paris were to dispose of it - by sale or transfer - as soon as my work in Roumania was completed. When it was decided to turn over a part of the hospital to the Serbs, I was asked to convoy the equipment to Belgrade - which gave me interresting drive I have written about...I thus had the good luck of driving all the way from the Black sea to the Adriatic, a most unique experience, and a splendid finale to my now rapidly ending military career.

August 1st, 1919 - Belgrade and the Car

For my services here in January, I was presented with the Serbian Red Cross medal, and at the same time proposed for the Serbian order of the Eagle; this letter will have to await the return of the Crown Prince, who is away just now. My hospital camion arrived late this afternoon, after a most strenuous and exciting journey, crashing through no fewer than four bridges; the worst accident delayed my drivers a whole day as the car had to be jacked up, inch by inch, while the rickety bridge was propped up underneath with beams and heavy boards. These Balkan roads are only built for ox-carts, very light ones at that. I was able to turn the camion over to the Serbians and get a receipt for it. In the evening, I attended a very interresting bridge party, given by the Dodges; all of the Diplomatic corps was present, and I again met Russian and Polish ministers who knew all about my family. It seemed strange to be addressed as Count Tarnowsky instead of plain "Doc" – the change seemed not unpleasant – under existing surroundings!

The following day, Aug. 2nd, I lunched with the A.R.C. officials, and in the late afternoon had an exciting adventure which, fortunately, terminated satisfactorily. Among the A.R.C. Doctors at Belgrade, is a Russian refugee, a female doctor who calls herself Dr Chumkoff, but who in reality is a Princess from Kieff. I had met her in January and found her very interresting, especially as she could give me much news of my family in Kieff and Tchernigoff. She - on this particular afternoon - asked me to visit the battle-field of Avala, about 20 K.M. from Belgrade. I accepted with pleasure and drove the car myself. Avala is a very high and steep hill, reached by a rather poor winding road, of very steep grade. My car is not very good on grades and has a nasty habit of getting overheated and stopping short at the most inopportune moment. At a very bad turning, near the top of Avala, the car lived up to it's reputation and stopped. I applied the brakes immediately, but the car slipped back several yards, and, rather than go over the steep edge, I had to swing one rear wheel into a ditch in order to stop it! It was then getting dark and beginning to rain. I waited half an hour for the engine to cool, and then tried to extricate the car, but could not do it. What to do now became quite a problem! We both shouted several times, but nobody answered - for the simple reason - as we found out later - that the place was quite deserted! Finally, I decided that it would be safer for Doctor Chumkoff to walk down the hill to where we had noticed a guard house and get help, while I remained to guard the car. I spent the next hour trying - without avail - to extricate myself and then, becoming worried over the long delay, I began walking down the hill myself, meeting two Serb soldiers on the way who had been sent up to help me. With a little digging and considerable pulling on the part of the willing soldiers, I was soon on my way down the hill, picking up the Doctoress on the way and giving the soldiers a substantial tip for their good work. It was nine o'clock when we arrived at the A.R.C. house; my driver grinned delightedly when he saw me drive up; we were to leave on the morrow, and he had been fearful lest I had hurt myself in some way. The doctoress, by the way, has quite reached the age of discretion! It might have been romantic - but wasn't in the least bit!

Precarious Burden of the Car

As on Sunday, there are no barges for Autos, except with the two P.M. Ferry, we could not depart before that time for our long drive to Triest. I must now digress a little and explain how it came about that I still had my car. It had originally belonged to the Auto-Chir, but my orders from the A.R.C. in Paris were to dispose of it - by sale or transfer - as soon as my work in Roumania was completed. When it was decided to turn over a part of the hospital to the Serbs, I was asked to convoy the equipment to Belgrade - which gave me interresting drive I have written about. At Belgrade, my chief mechanician, Thorpe, was taken over by the Serbian commission as chief of their transportation service, and the first thing he discovered was that a lot of spare parts were lacking which could only be purchased in Italy. Hence he decided to go to Triest with a camion and my car. This just suited me to a T as, with a camion full of extra gazoline and oil, I could make the trip without much danger of a serious breakdown. I thus had the good luck of driving all the way from the Black sea to the Adriatic, a most unique experience, and a splendid finale to my now rapidly ending military career. So, with the remnant of my unit-four men - a camion and my car, we sailed up the Saave to Semlin, disembarked and began our trip up the picturesque Saave valley. The roads were fair for the first thirty K.M. to the town of Indjii; from there to Ruma, a distance of twenty K.M., they were abominable in the extreme. I ran my car in a ditch, trying to avoid a swampy bit; we had no sooner extricated it by jacking it up and pushing it over the hump, than the camion drove up, and, in spite of our warnings, tried the swamp the result being that we had to drag it out by hitching a heavy rope to my car and pulling it out backwards! This delayed us several hours, so we decided to drive on as far as possible before putting up for the night. We had an ample supply of canned food with us, but the men had forgot-en to get bread in Belgrade, so we bought some in the little village of Tovarnic. The people spoke German in that village; in most of the others, Roumanian or Serbian. The country was very prosperous looking, everywhere splendid crops of wheat being harvested and wonderful corn fields with promise of a huge harvest in the near future. The quantity of cattle, pigs, chickens, ducks, geese, etc, was simply phenomenal; every village we passed was full of animal life, and it was hard not to kill any of the tame fowl or animals; they were totally unaccustomed to an Auto, and had no idea how to guage rapidity of movement, waiting in the middle of the road until the car was almost on them, before they would deign to get out of the way! My driver finally hit upon the plan of sitting up in front with me, with my walking cane in his hand, and whacking at the pigs and dogs as we passed them by! It was most effective; the geese showed the most sense of all, got out of the way in time and hissed us loudly as we passed! Using our head lights, we were able to drive until ten P.M., stopping near a well and large wheat field where the nice dry wheat was still piled up in convenient stacks, under which we could sleep. We each had two blankets apiece; the four men arranged their beds out in the field, while I curled up in the auto - after placing my revolver in a convenient pocket! Several groups of peasants, and a caravan of gypsies passed along the road and looked in on me, but nobody tried to molest us. We had barely fallen asleep when a terrific storm broke, compelling the men to retreat from the field to the car and camion. We were all so crowded that none of us got much sleep during the balance of the night. By four A.M. everybody was so cramped and stiff that we were glad to boil our coffee, eat a good breakfast, and make an early start. It was still drizzling, and the roads were heavy and slippery; the best we could do was about ten miles an hour - not always that. Throughout the villages of Vukovar, Nastar, Vincovice, Brode, Oriovatz, Neu-Kapel, Neu-Gradiska, Raiich, and Novska, the inhabitants were very hostile towards us – probably because we were strangers! I had an American flag glued onto my windshield, but these primitive people had probably never seen one before, and it meant nothing to them. At the small village of Banovicryza, I stopped my car to await the arrival of the camion; a peasant ran up to my car and whacked at me with a heavy cane, missing my head by inches! I whipped out my revolver and shouted imprecations in four languages, but he slunked away and the rest of the inhabitants did not molest us! I ordered the camion to keep in close touch with the car, so as not to become separated, and pushed on as fast as the condition of the roads would permit. At the very next village (Novska), the camion developed engine trouble and came to a full stop! In two minutes, the entire village was surrounding us, asking us who we were! I explained – in German – that we were Amerikanskis – whereupon one of the men spoke up in broken English, telling me that he had worked in the coal mines of Illinois. "How you like Ziegler?", said I, luckily guessing the right place. "Die sind virklich Americaner" said the man, turning to his fellow-villagers! From that time on, they merely looked on with curiosity while the engine was being put in running order. Off we started again, arriving at the decent sized village of Ku-tina at half past nine. There we found the majority of the inhabitants to be of Germanic extraction, and it was with a genuine sigh of satisfaction that we drove our cars into the courtyard of a clean looking Gasthaus and secured three good clean rooms for the four men and myself. As soon as our papers had been viséed, the word was passed around that a party of Americans had arrived, and the Burgomeister, doctor, priest, and banker all came to pay their respects to me! The old village doctor had studied with Professor Von-Eiselberg in Vienna, so I was able to discuss the old Allgemeines Krankenhaus with him – having been there myself in '13. After a late supper, we all turned in and had a glorious night's sleep, the beds being spottlessly clean and fresh. In the morning, all of the village notables were on hand to wish us "eine gluckliche reise", one of the men giving me a letter to mail to a relative of his in Akron, Ohio.

A Scenic Drive

From Kutina on, we were in Croatia; the difference was most marked; there certainly was a gap of three centuries of civilization. The farms were models of neatness; the villages devoid of pigs, chickens, and geese running around loose; the houses all carefully built and artistically painted – even to the roofs – and the inhabitants most polite. Unconsciously, I put my revolver away in my bag and felt perfectly safe! We pass through Agram (or Zagreb), the capital, at noon. I have already described it in my January letters; it had taken on new life and was a modern, up to date city in every sense of the word. We lunched along the banks of the Saave, near a delicious cool spring. It was a religious holiday, and all of the peasants were evidently going to a picnic, as quantities of them passed us on the road, men and women alike wearing white linen clothes, the bottom of the men's loose trowsers having a fringe all the world like a Turkish towel, the upper garment being a long shirt falling to the knees, tucked in at the waist by means of a very handsome brass studded belt of brown leather; a felt hat with a flower or feather in it, and black leather top boots, completed the picturesque costume. The women's waists and skirts were most wonderfully embroidered, the hair neatly platted and tied with red or ping silk ribbon; most of the women carried their top boots in their hands, their stockings sitting comfortably on top of the lunch basket which they all balanced on their heads. A group of very pretty girls posed for me while I snapped them with my Kodak. All looked happy, contented, and well-to-do; the children well fed and healthy looking, the men distinctly handsomer than the women. The Austrians tell you that the Croats were not loyal during the war; that they refused to share of their plenty with famine-stricken Vienna and hid their produce where the officials could not find them – "ca, c'est une autre histoire!".

Leaving the Saave valley, we began a drive of 160 K.M., which, for beauty and variety of scenery, I have never seen equalled. The magnificent national highway passed up and down dale through magnificent fir-tree forests, with clumps of evergreens, maiden-hair ferns, and wild flowers to the right and left, beautiful green grass everywhere, and a profusion of wild birds and hare such as would make a hunters sojourn there like a touch of real paradise! One silly young hare actually ran in front of my car for several miles, apparently enjoying the race; he finally darted into a thicket and sat down to think it all over! It was simply ideal, in spite of the fact that we had to put up the top on account of short showers. The lack of sun prevented me from taking any pictures, but the scenery is indelibly impressed on my mind. We arrived at Laibach (Llubianja it has now to be called), at half past seven in the evening. I tested my bump of locality by driving my car right up to the door of the Union Hotel where I had spent a few hours in January. The manager told me that he had no rooms – so I drove to the military commandant and had two rooms requisitioned at the same place! The hotels prefer to cater to civilians as they will pay any price for a room; the manager thought he had gotten rid of me – and was correspondingly crestfallen when I marched back with the requisitions in my hand! A requisitioned room costs about a quarter of the ordinary rate; so he lost out by being too grasping! He had attempted the same trick on us in January – with the same result; I took pleasure in reminding him of this fact – after which he was most obsequious! Food prices had dropped fully fifty per cent since January; as the value of the Kroner had dropped to one fifth of the value of the franc, and you could get seven of the latter for a dollar, a meal costing 35 Kroner represented exactly a dollar meal! At Kutina, five beds, beer, and eggs for four people cost just two dollars! Now is the golden time for American tourism in Europe. When I figure out my colonel's pay in Kroner, I feel like the whole Anheuser-Busch corporation! The Serbian army is in full control of both Zagreb and Llubianja – the Croats are beginning to scratch their heads and wonder whether they have put their money on the right horse – after all! It is all a bad muddle.

Leaving Llubianja on the morning of August 5th, we again had a lovely drive over the Julian Alps. The scenery was wilder, but most picturesque, the road a marvel of scientific engineering. At Longatico ended the present Jugo-Slav zone of occupation. We were halted several times by finding massive swinging beams across the road, which stopped us until the guard came out and looked at our papers. At the Italian confine, the preparations against possible surprise were most effective; not only was the road barred, but there was a heavy guard, with machine guns sweeping every corner of the road. There was a formidable concentration of troops for miles and miles along the main and all side roads. We were not popular as we drove along; "Via di qua, porcheria", seemed the favorite expression of the soldiers as they recognised flags and uniforms! We were hailed by two Italian officers who were stuck on the road by a puncture and no extra tires on hand; they gratefully accepted my last spare inner tube – which I gave them in memory of the good old days when the war was in full swing, and Europe exhibited pathetic gratitude for our moral, physical, and material help! Those days have gone by now; we are former honored guests who have had to stay on too long!

I had hoped to be able to visit the famous grottoes of Adelsberg, but the Italian zone of occupation made it necessary to get a special permit and wait until two P.M., as it was only eleven o'clock when we passed the entrance to the grottoes, I pushed on, especially as I did not know how soon I could get a sleeper to Paris.

Crossing the famous Carso plateau, I saw the most severe evidences of the Austro-Italian war; the effects of the bombardment were almost comparable to those on the Franco-Belgian front. We arrived at Triest at five o'clock and secured rooms at the Milano (ex. Volpich, with the statue of the late empress removed, but the pedestal still in place with "Elizabetta" still carved upon it!) and put our cars in the U.S. naval garage. There are still several U.S. destroyers at Triest and Fiume; the naval officers I spoke to seemed to think that they would have to police the Dalmation coast for several years. There had been Bolchevistic riots in Triest two days before; we were told that processions of children had been coached to cry "Viva il Bolchevismo", whereupon the soldiers shot into the crowd, killing and wounding several children and women. The city was still under martial law when we arrived, armed soldiers patrolling the streets night and day, no meetings being allowed. While my men were having a soda at the local Y.M.C.A., a guard of U.S. marines came running in and ordered everybody to remain indoors until it was safe to venture out again!

These notes have been written on the Simplon-Orient express. I have lifted my eyes occasionally to gaze a fond farewell on Venice, Padua, Vicenza, Verona, and Brescia; we are now nearing Milan. I am sorry that we cross Switzerland at night-time, but we are due in Paris tomorrow morning, August 9th. If my good friend Col. McAffee in Paris has kept his word, my home papers ought to be awaiting me there! Pardon mistakes; the train is a fast one, and it is hard not to hit the wrong letter at times!

Donna e Mobile

Letter dated July 16, 1919

My dear young lady, permit one old enough to be your father the liberty of calling your attention to the futility of your questions - your environment, education, and traditions are so different from ours that it would be impossible for you to see our viewpoint.

Donna e mobile

It was no less a personage than gay, débonnair King Francis the First of France and Navarre who once exclaimed:

" Souvent femme varie
"Bien fol qui s'y fie!"
"Woman often changes; quite mad is he who trusts her!"

Of course, this is a base and utterly unwarranted calumny on the fair sex; man alone is supposed to be floppy and fickle; nevertheless I found myself whistling Verdi's famous aria from Rigoletto and quoting – for my own special benefit – François de Navarre's quaint poetry, in the language of the period, after opening and reading the letter which had been delivered by special messenger at my hotel on the Callea Vittoria, Bucharest, Roumania!

"My dear Colonel:

"I saw you yesterday, driving with YOUR FRIENDS. The visit to the Roumanian country-house is cancelled – those who were to have been our hosts have already been notified of the change in our plans.

"We shall never meet again in this world, nor correspond with one another, so that we shall have parted the best of friends!

M.D.

Bucharest, July 16th,1919."

The friends referred to in the extraordinary letter were my former travelling companions on the S.S. Imperatul Traian bound from Marseilles, France to Costanza, Roumania, the Marghilomanist ex-secretary of Finances and his wife whom I had called upon as an act of polite formality and taken for a drive along the fashionable Chaussée - where everybody in Bucharest who thinks he is anybody drives an afternoon to see and be seen. For the benefit of the "non-conoscenti", it might be well to explain that Marghiloman, a prominent Roumanian statesman prior to his country's participation in the World War on the side of the Allies, had been an ardent Germanophile - or Centrophile as they were called in the Balkans -. He and his followers were therefore in total eclipse, politically, after August 1916. With the very characteristic intensity which distinguishes political opinions in the Balkan countries, a Centrophile became more contemptible in the eyes of the Ententophiles, or adherents of the Allies, than, shall we say, a Mugwump in the early post-Civil-War years in our own country?! It was true that, after Roumania had been conquered, the Marghilomanists were able to softly murmur: "I told you so!" but, as soon as the Allied star again became ascendant, they once more found it convenient to take urgent trips which took them out of their country!

The letter itself, written in French, had been sent to me by a young Roumanian girl, also a fellow-passenger on the same steamer. The World War had finally ended, and poor bleeding Europe was staggering hither and thither as though in a drunken stupor, picking up the broken threads of her once peaceful life one by one, hesitatingly, haltingly, suspiciously. Having closed my Evacuation hospital in France and seen the various smaller units which had composed it's staff, scattered in as many Base Hospitals preparatory to being shipped home, I had been entrusted with a special mission which took me to the Balkans. With my staff car and driver, I had finally arrived in Bucharest, Roumania.

The Golden Horn, Constantinople.

While having vaguely noticed her on the promenade deck or in the music room of the steamer, we had not been assigned to the same table in the dining room and no opportunity for an introduction had presented itself or been sought. It was while sitting on the upper-deck - we had anchored in the harbor of Constantinople just opposite the Golden Horn and, after a hectic day of sight-seeing, I had planned a quiet evening of contemplative meditation - that a volume on Leonardo-da-Vinci, written by a Frenchman, caught my eye as it lay, unopened, on a nearby steamer chair. Idle curiosity prompted me to pick it up and read a few passages, here and there, haphazard, as fancy dictated. Becoming engrossed in it's idealistic depiction of one of the Renaissance period's most famous men, I was not conscious of the fact that a young girl had been promenading the deck near me until she stopped in front of my chair and, pointing to the book on my lap said, in French: "I am glad to see that Monsieur le Colonel takes an interest in my book; Leonardo is my beau-ideal, almost my God!"

Arising and profusely apologising for my absent-mindedness in picking up and reading a book which did not belong to me - I really had thought that it had been borrowed from the ship's modest library and was therefore almost common property, I immediately offered to return the da-Vinci volume to it's rightful owner.

"No. no, Monsieur, please keep and read it from first to last page; I have finished it - in fact I have read it several times over. As a punishment - which I assure you is undeserved - you may discuss it with me on some future occasion!"

Thanking her for her kind suggestion, I started to introduce myself, but was again interrupted.

"I know all about you, my dear Colonel; you were very kind to my dear Uncle Mr L....., the Roumanian representative at Marseilles, and he pointed you out to me just before our steamer sailed. I am his niece, Maria, going back to my beloved Roumania with my mother and cousin, after studying Art in France for a year."

I smiled inwardly at the recollection of the excitable little man - a blending of Polish, Greek, and Roumanian blood - who, though so hard to pin down to concrete statements or decisions had, nevertheless probably done the very best he could for me in his semi-Oriental, utterly un-American, un-business-like way. Having called one day at his Marseilles office to get final sailing instructions and assurances that my staff-car could and would be taken aboard, he had beseeched me to protect him from the uncalled for threats of violence which a young American employee of the Near-East Relief Society was, at that precise moment, making against him. Belonging to the "mucker" class - one of the many misfits who disgraced America during and after the War, the roughneck had sought to camouflage his utter ignorance of foreign languages and customs by means of bullying gestures and water-front profanity. A few terse military commands soon put the arrogant youngster in his proper place!

Petite, almost of the scrawny type of young girl, Mademoiselle Maria was neither pretty nor beautiful - yet her face was an interresting one to gaze at. Black hair, almost swarthy olive complexion, expressive brown eyes, and a rather large mouth, offset by a double row of perfect teeth, her features in repose were not in the least bit striking - but the minute she spoke, her whole appearance changed. Looks and speech alike denoted sincerity, seriousness, and an utter absence of frivolity rather unique in one so young. Dressed indifferently - not the studied indifference of the sartorial and pictorial artist - but the genuine carelessness in dress of one who depends on intellect and not trappings to force recongition; face and hands absolutely innocent of the tell-tale signs of beauty-parlor ministrations she would, in all probability, be classed as queer or freaky-looking by a majority of her less intellectually endowed sex. To me, she was most refreshing, a novelty - one of the "new-woman" type of Europe and, our mutual status having been clearly defined, it was not necessary to be on one's guard. The World War was over, and my particular mission was of a purely philanthropic nature, with nothing to hide.

"Are you Americans happy?" The abrupt question was quite startling, but I was given no chance for an immediate affirmative or negative answer - my new acquaintance belonged to that peculiar species of human who can, when so minded, carry on a conversation on any topic without bothering to pause for answers or even comments from their audience!

"I mean", she continued, "that Americans, the few whom I have seen in Marseilles and Bucharest, always seem to have such a preoccupied air; one saw them rushing about the streets, driving rapidly, eating rapidly in restaurants, drinking feverishly in cafés, running from the theater as the curtain descended on the last act - always in a hurry, always looking anxious - have you then no time for rest, relaxation, and the quiet enjoyment of the beautiful things in life?"

"But, Mademoiselle!" I began feebly......

You accomplish miracles!
Crossing the Danube on an improvised raft.

"Yes, yes, I know! You accomplish miracles, you are perfect dynamos of energy - you are forever pursuing something beyond you - what? Have you a goal; if so, when do you - when can you....attain it? We, of old Europe - I class myself as one of them, for we Roumanians are Rumeni - ancient Dacian Romans - we Europeans are born old; we develope in an atmosphere of retirement, of introspection, of ancient Graeco-Roman philosophy, which early teaches us to regard life as but a fleeting moment of perception, of feeling - to be followed by darkness!

"Thanks to this background, we encourage the joy of living and loving - synonymous terms to many of us - you young Material Americans do not seem to understand the real inner or outer meaning of life; at least not to our way of thinking!"

"Really, Mademoiselle, you do us an injustice; we do enjoy living, we do pause - in what you probably term our mad career - long enough to...."

"Certainly, mon Colonel, certainly - you breathe and eat and move about - presumably you must sleep, sometimes, therefore you can be said to live; you pause long enough in your frantic rush to found a family - just as you probably acquire a house and furniture - regarding all equally as necessary details........but they are, to you.....details - not life - not love! With us, Monsieur ah!, with us........it is different, very different!"

"Perhaps you do not like us, Mademoiselle; I am sorry..I had hoped that we had been of some comfort to your people.."

Tears sprang from her dark flashing eyes as she - once more - checked me.

Waiting for the daily distribution of food & clothing.
American Red-Cross activities in Roumania.

"Pardon, mon Colonel, pardon! I do like you....I love Americans! Can I ever forget the soup-kitchens in the streets of Bucharest, the distribution of clothes to the half-naked peasants, the hospitals, dispensaries, and other monumental activities of your wonderful, marvellous, life-saving American Red Cross? We Romanians can never, never forget your help; it was efficient, coordinated, complete and yet......Oh! how can I make you understand my meaning? You, with your Slav name and blood - although you are an American officer - must have some of that old strain in your veins....you must know, must sense what I mean!

"Your services to us - wonderful though they were - lacked.... spirituality.... you gave abundantly, lavishly, at times wastefully of your material possessions, but of yourselves.....you gave so little! Master minds directed your organizations, level heads worked out details in all of it's ramifications....but the heart....it was not there! You seem...so...cold...so undemonstrative - is that what you mean by the terms Superman and Superwoman....a complete crushing of the senses?"

"Mademoiselle," I answered, "we rather pride ourselves on being addicted to sudden, at times uncalled-for, or incomprehensible outbursts of sentimentality, and for that reason..."

"Sentimentality! Yes, Monsieur le Colonel, I agree with you; you must be a sentimental race...at times...but sentiment...is quite another feeling, profound, satisfying - governed by both head and heart! Are your people quite certain that their - as you term them yourself - outbursts of sentimentality are genuine; is it not possible that.... may there not be a varying amount of affectation in what you overdo...on the impulse of the moment?"

"Impulse, my dear Mademoiselle, is perhaps the right term to apply to young America; is it not one of the classical attributes of youth - individual or nation - it's very much the same in the aggregate!

"For that reason, your estimate of us is perhaps a fairly correct one, as far as it goes. Adolescence is given to unthinking outbursts of enthousiasm, jingoism, sentimentalism, or...puppy love; it's wonderful, sublime, and inspiring....while it lasts, but our very youth prevents us from feeling very deeply - or consistently.

"I really understand what you mean when, in spite of our dynamic energy, you find us...cold and undemonstrative!

"There is often a something lacking in our spirituality, a vague, intangible, elusive spark which is missing. For many years, I have sought an answer to this riddle, have endeavored to apply words which might describe my meaning, that my friends might know what I meant, how I felt - why I felt it was absent.

"It isn't lack of personality nor of magnetism - we are a magnetic race, and our population contains in it's midst a proportion of personalities equal to that of any other country on earth - it is perhaps....it may be materialism - using this now much perverted term as typifying our national pursuit of the purely physical results of modern energy - and surely youth demands little else....at first!

"You see, Mademoiselle, you have got me back to my starting point....we are young; is it wise....were it better for us - and for you - that we grow old?"

How would my youthful inquisitor - she may have been twenty-four-or-five years old, not more - take my last challenging question? It opened up an infinite vista of philosophy and might lead....whither?

Intellectual conversations, carried out by members of opposite sexes in polite society are all too often barren of results, because one cannot delve to any considerable depth without reaching the point where naked words need garments to shield their truths and evasion or subterfuge replaces honest facts or accepted opinions!

With our innate hypocricy - which we often erroneously term good-breeding, delicacy, pure-mindedness...and so forth..., we avoid or sidetrack the really important part of many discussions because, forsooth, the subject begins to tread on what we complacently term "dangerous ground"!

I know just about how far conversations on philosophic topics can be carried on with my own countrywomen; I can sense, quite instinctively, the limit beyond which it is forbidden to go, but this cosmopolitan damsel was somewhat of an enigma to me and, while it was probable that her apparently insatiable curiosity was quite innocently impersonal, "watch your step" and "safety first" very naturally came to my mind! She, I mentally decided, would have to do the leading,....I would follow, warily! No real gentleman 1ikes to inflict a moral shock or injury on an immature mind - but my questioner had expressed opinions with all the apparent assurance, which one associates with a mellowed experience of life's problems. How much honest interchange of ideas did she really desire?

Turning suddenly from the deck-rail - from which point of vantage she had been gazing at the Stamboul lights - thus allowing me the necessary moments of reflection-, she took up the thread of our conversation where I had left off. I noticed however that she did not take up my challenging question!

"It must be materialism which accounts for your immature mentality - how else could it be safe for your American girls to enjoy such complete liberty without a total breakdown of the codes and rules of conduct on which our white civilization rests? I have read much about your women and girls; I have seen them in Europe, and I long for the untramelled freedom which they enjoy throughout their lives - be they married or bachelor-women...you see, Monsieur, I even know that there are no old maids in your great America, only bachelor-women....., but I have never been able to understand the...pardon my utter frankness mon cher Colonel...sexless camaraderie which apparently is the rule in both city and country life in America! With us, in Europe, the...temptation would be too great; with your young men and girls....there seems to be none - are you then so utterly devoid of....feelings? What is it in the American, in the Anglo-Saxon, that makes unnecessary the constant surveillance, the chaperonage, the odious duennas - all the protection we are subjected to from the day we lengthen our skirts and do up our hair? Is it because your boys and girls have set up for themselves a certain material goal.... financial or social..that they can sternly refuse the...cup of life, even when almost placed at their lips - until they have attained that material heaven which they seek - or are you really devoid of...feelings? Is love merely a civic duty with you, or is it life itself...as with us? In your frantic rush after the "Allmighty Dollar" - she quoted our own favorite expression with a deliciously quaint little accent all her own - "are you not missing something which you might quaff of if you paused long enough to satisfy your thirst....providing you have any?"

"My dear young lady, permit one old enough to be your father the liberty of calling your attention to the futility of your questions - your environment, education, and traditions are so different from ours that it would be impossible for you to see our viewpoint. Our system of education enables the sexless camaraderie - as you perhaps not inaptly term it - to be a fact - with occasional exceptions of course; imitation of the same intermingling of sexes in Europe in an effort to live "a l'Américaine" has, it seems to me at least, been rather disastrous...so far... because background, customs, and...temperament, are so different. Perhaps it is the games and athletic sports which we indulge in as boys and girls, the eager desire to win in competitive physical tests...which enable our youths to curb their natural desire for sweets, frivolities, late hours... anything which might hamper their maximum physical development; perhaps the hope, the vision of such innocent material triumphs does tend to keep us...immature!

"We do think much more about the health of our bodies...and minds...than you do; the average American is very apt to brush aside anything, anybody...who might interfere with what he - rightly or wrongly - considers to be success in life, and he is able to do this very thing rather easily because of his early discipline, his early acquisition of the power of physical control and mental mastery over self...perhaps also because his whole being is not overflowing with...temperament! And yet, when he has either attained success - or it seems within his grasp - this same materialistic American man will give his all to the girl he has chosen - or who has chosen him - with a generosity unequalled in any other country in the world!"

"Will give his all, Monsieur - you mean his money, lands, houses - all the material evidences of his power to accumulate dollars - but what of...love? Tell me, you who know men on both sides of the Atlantic - how do they compare, your men and ours, spiritually, emotionally, temperamentally? I have heard it said that American men make wonderful husbands but...indifferent lovers - is it true?"

"How can I - a man - answer your question?"

"Bravo! mon Colonel is quick-witted, and I am caught in my own trap! What I really wanted to know was this: Can a man who has spent almost all of his adolescent life - we are not considering hours of study which, by the way, seem much shorter and less strenuous than with us - in games requiring team-work, self-repression, physical and mental discipline - and so forth -, can this man, when he decides to marry.... reawaken sensations which he has forced to remain dormant... or which have never yet appeared in him - will he not tend to remain an athlete - industrial, commercial, or financial - the same dynamo of energy which arouses our admiration and...astonishment? Has he time or natural inclination for...anything else....but to accumulate more and more material riches, in order to pour them on his queen's lap? And she - is she not....at times...hungry for more spirituality, more love?"

"Perhaps, Mademoiselle, the fact that in America - where young men and girls choose their mates with little or no supervision or interference on the part of their respective parents - we have more divorces granted each year than in all of Europe, might seem to substantiate the reasonableness of your question; might seem to prove that where there is less love there is also less mutual forgiveness in marital life.

"American women who have had matrimonial experiences on both sides of the Atlantic tell me frankly that, after all, their own countrymen are....less thrilling but....safer! Does this mean that the wife and mother who remains steadfast to her own vows prefers safety to...thrills - or does she ignore the latter meaning, never having experienced them herself? We have, in the English language, an old adage which states that: "when ignorance is bliss, t'is folly to be wise"; it may apply to the situation we are discussing - and it may not - I really don't know; one can only surmise such things. But, my dear young lady, you are really getting me into deep water; fortunately, I am an excellent swimmer!

"Life, as I understand it, cannot avoid being shipwrecked without self-discipline - mental as well as physical -; what we call civilization is the result of the sum total of individual sacrifices, repressions... suppressions if you wish - of practically all of our animal instincts for the good of society at large. Perhaps our Anglo-Saxon discipline is of a sterner variety than is yours in Europe, because we are so young; I sometimes think that we also are getting older. Is your philosophy - which you intimate so cleverly - nearer the truth...nearer life...than is ours?"

"Qui s'excuse, s'accuse" - he who makes excuses accuses himself! I almost think, mon Colonel, that you incline to my way of thinking? No!...don't deny....or affirm - it would spoil the evening for both of us, and I have so enjoyed this heart-to-heart talk with you! Do not take it as a personal compliment to you - unless you so desire - if I confess that it would not have been possible for me to discuss such.... intimate subjects with any countryman of mine; he would have regarded the orientation of my questions in the light of.... an invitation! It could not have been!

"I shall probably never have the opportunity of knowing the American man or woman in America, but I do want you to know us as we are - chez nous, in Roumania. Are you, will you be free to accept invitations?"

"Quite free, Mademoiselle, I assure you. Just as soon as I shall have delivered the special Field Hospital which we call an 'Auto-Chir' to her Majesty Queen Marie, my mission ends and I can either rush back 'a l'Américaine', or linger a few days or weeks by the wayside as fancy dictates."

"How splendid; how fortunate! My plan will work out beautifully. I shall speak to dear friends of mine in Galatz; you will know Roumanian life in the country, on a real Boyar estate. The visit will be as long or as short as you wish to make it; it will give you ample opportunity to study native types - including myself!"

Laughingly, she added: "You young stoics of America, when in Europe.....change your philosophy rather easily...at times! We are probably all of us mortals very much alike, but you of young America must have a passion for hiding your inner thoughts...your inner life....from indiscreet eyes?!

"See, the lights of Stamboul are all out - it must be frightfully late...or early! Fortunately both my mother and cousin are sound sleepers, otherwise....Oh! la! la!

"Bonsoir, mon cher Colonel, and pleasant dreams!"

"Good-night, Mademoiselle Maria - and thank you for the most truthful and candid discussion I have ever had with a member of the opposite sex on so...delicate a subject!"

"We did avoid the rocks, didn't we, Monsieur?

"Admirably!

Bonsoir!

Bonsoir!"

What an extraordinary type! Of course, a mere adventuress might conceivably have opened her campaign in a similar fashion, but this high-strung, gifted young girl of eminently respectable parentage - the mother a demure refined-looking widow of fifty years of age or thereabouts, the cousin both charming and pretty, and an Uncle occupying an important government position - established a new precedent for me and excited my curiosity. She was worth further study, if opportunity offered.

The following morning, having satisfied all international formalities, the Imperatul Traian was allowed to continue on it's voyage. The unsurpassed loveliness of the Bosphorus - or Hellespont of the ancient Greeks - kept me rushing from starboard to port in order not to miss seeing the many marble palaces of Terapia or the rugged beauty of Leander's tower. Two hours after weighing anchor, we entered the Black Sea. The drab monotony of the marshy coast-line soon put an end to further sight-seeing; I sought and found Mademoiselle Maria on the promenade deck. Was it because of daylight and absence of full-moon, or fear of indiscreet ears? The previous evening's spell was broken! It seemed impossible for my young fellow-traveller to resume her psychological and physiological train of thoughts - as for me, it was a relief to converse on....safer if less exciting topics!

On the subject of racial, religious, and political problems pertaining to Roumania, she was just as intense, as convinced - if not convincing - as she had been on the previous evening! Quite evidently an omnivorous reader - though perhaps of indifferent digestion - she, in common with many older men and women of the Balkans, only remembered periods and events which redounded to the credit or glory of her country; after all, are we any different? Do we care to discuss our treatment of the Indians, of the Negroes, amongst strangers? Throughout the World War, her sympathies had all been on the side of the Allies and more especially of France. She hated Centrophiles, loathed the very name of Marghiloman, and refused to converse with the latter's ex-minister of Finance or his wife! Loathing the Greeks, she bitterly regretted having such a strong sprinkling of Hellenic blood in her veins - perhaps some ancestor of hers had been one of those shrewd overseers - a Phanariot from Constantinople - who had ground taxes from the poor lower Danubian peasants, for the benefit of a Turkish Vizir?

As befitted an army officer of a friendly nation, it was impossible for me to discuss political situations - I could only listen sympathetically without expressing opinions or making comments which might easily have hurt the feelings of some Roumanian, Greek, or Bulgar passenger who had been kind enough to trust me with his own private version of the Balkan tangle! Ententophiles, Centrophiles, Anglophobes and Frankophiles, Anglophiles and Frankophobes; pro-this and pro-that - all had, turn-and-turn-about, expounded their theories of national aggrandisement to me under cover of such seclusion as a corner of the lower deck offered; all had shown the same brand of intolerance towards other convictions or beliefs than their own individual ones which we - half jestingly it is true - associate with Indiana politics of the late nineties!

In spite of repeated attempts to lead me on, I dodged every issue and allowed Mademoiselle Maria to unburden, for my special benefit, her very strikingly personal view of the Balkan imbroglios. Again and again, her opinions had all the assurance and instability of extreme youth - she hated with an intensity, which was quite comical. Many of her fellow-countrymen on board steamer did not - in her estimation - deserve hanging; boiling oil or the rack would have been used to remove them from office....if she could but have had her own way!

On the morning of our glorious Fourth of July, the steamer entered the small but quite modern port of Costanza, export center of the Roumanian oil indistry on the Black Sea. Being quite naturally curious to see the effect of warfare in

Roumania, in order to estimate it's destructiveness as compared with that of the battered western front, I landed as soon as shipping formalities were over.

Accompanied by the Roumanian ex-official and his young bride - who intended reaching their estates by train - we drove around what they considered to be the most modern and progressive of their cities. Formerly a typical nouveau-riche excrescence with gaudy villas and palaces of nondescript architecture surrounded on all sides by forlorn-looking one or two story buildings of native style, Costanza was, in July 1919, almost a dead city. A few of the larger buildings and the new Casino on the sea-front had been hit by enemy aerial torpedoes, but systematic looting accounted for most of the dammage inflicted on the inhabitants. Tile-roofs, doors, windows, staircases, gutters, rain-pipes, and kitchen-drains, all had been removed and taken out of the country, principally - I was informed - by Bulgars and Hungarians. Of furniture, little or none was left.

Nowhere was there evidence of any concerted or efficient resistance; no trenches, strong-points, or remains of barricades could be seen. It was sadly evident that the short, decisive conflict had been entirely one-sided. The entrance into the War of the unprepared, badly trained and poorly led Roumanian army and fatal delay in the promissed delivery of munitions and other war material constitute one of the real tragedies of the latter months of 1916! The rapid overrunning of three-fourths of the little country by the combined armies - Austro-German-Bulgar - under generals von-Mackensen and von-Falkenhayn must have been as easy an affair as the scattering of an ant-hill with one's foot - the comparison is not as exaggerated as it might seem, nor does it necessarily convey a reflection on the relative fighting ability and qualities of the two armies; after all, one simply cannot fight a modern campaign without modern equipment and a knowledge of modern tactics - Roumanian officers frankly admitted they were woefully deficient in both!

We proceded up the Sulina canal

A glorious swim in the sea, at a wonderful sandy beach called Mamaia, sharpened our appetites. My Roumanian cicerones insisted on my lunching with them in one of the few restaurants which had reopened after the Armistice.

In spite of the evident misery of the peasant classes, the quantity and quality of the foods and wines served us were excellent - in fact, lunch turned out to be a banquet of semi-oriental splendor! Russian influence was immediately manifest; mountains of Zakouska - or hors-d'oeuvres - were passed around, to be washed down with Mastica, the Greek and Zuika, the Roumanian national beverage - both reminiscent of Vodka, only more fiery!

It was truly a feast of endurance; when it was all over and my hosts had departed for the railway station, I really welcomed the mile long walk back to the steamer...it was quite necessary!

We sailed that same afternoon and proceeded some hours later up the Danube river - or rather the Sulina canal, which enables shipping to avoid it's treacherous and changing Delta - to the town of Galatz. Mademoiselle Maria did not make her appearance on deck, nor did she even appear after supper in the music room, where the handful of Americans monopolized the piano and sang patriotic airs with the forgetfulness of words for which we are, alas!...quite famous!

It was not until the following morning, and while I was watching the unloading of my staff-car preparatory to it's trans-shipment to the small ferry-boat which runs up the now shallow Danube as far as the city of Braila, that my young enigma reappeared upon the scene.

"Good morning, mon cher Colonel; I hear you intend driving in your car to Bucharest? I happen to be quite familiar with the roads; don't you want to take me along as your guide and interpreter?"

"You are most kind, Mademoiselle, but - my car is now using it's last and only set of tires, and they are pretty well worn out. With several pieces of luggage and my driver, I don't see how there can possibly be room for three ladies!"

"Oh! but mother and cousin Elizabetta go by train -I offer you my services.....not those of my entire family!"

My driver had warned me in Marseilles that the tires were rapidly reaching the blow-out state; in spite of frantic efforts to replenish our supply, it had been impossible to obtain the right sizes, and we therefore depended on the French Quartermaster Depot of the army of the Orient in Bucharest to requisition the necessary replacements. Until we reached the Roumanian capital, we could only trust that good-luck and a light load might combine to carry us through safely. The unfortunate situation had already been explained to the three American Red-Cross nurses on board steamer who - naturally - had hoped to be able to complete the journey with me.

"I regret it exceedingly, Mademoiselle Maria, but it would be dangerous to risk taking a lady with us. My military driver and I are accustomed to rough-and-tumble situations, but you......what could we do with you, should we have to abandon my car?"

"Do! why do with me what you would do for yourselves, of course! I will gladly take - what you call it in America - one....sporting chance with you!"

"No. Mademoiselle; I do not wish to appear discourteous or arbitrary. I wish I could accept your offer, really I do but.....you see...it is utterly impossible!"

"Is that your last word, Monsieur?"

"Alas! yes!"

"Monsieur le Colonel does not, I hope, forget that - even in Roumania - it is not man who generally has the..... last word!"

With a stiff little bow and an enigmatic smile on her lips, she turned on her heel and disappeared. Whew! what a very determined young person she was!

With true semi-oriental shiftlessness, my car did not reach the ferry-boat landing until the steamer had started on it's first trip upstream. One soon learns not to loose one's temper in the near-East; it does no good and only spoils the digestion!

Four in the afternoon being the next scheduled starting hour, I retraced my steps and sat on the deck of the Imperatul Traian with the laudable intention of bringing my diary up to date. Barely had my thoughts been collected when up came the Captain, accompanied by no less a personnage than the Director-General of Roumanian navigation who - almost tearfully - begged me to take him with me to Bucharest where he had, so he said, very important government affairs to attend to. Why had he not caught the train; why had my car missed the first ferry-boat; what had Mademoiselle Maria to do with this new entanglement? Having refused the use of my car to three American nurses and to Miss Temperament - how could I take this official with me?

I did!.....largely because, having explained my mission and the expected arrival of my Field Hospital, packed in it's many trucks, this official immediately offered to provide the necessary flat-cars for their transportation from Galatz to Bucharest. Be it here written, as a matter of record, that he did keep his promise!

"Excellency," I laughingly added, as a precaution against possible future misunderstandings, "I only make one stipulation - that you promise not to 'blow-up' if we have a 'blow-out'! He assured me of his willingness to take a chance on tire trouble and, to show his gratitude, immediately got up a special lunch on board a sister ship at anchor nearby, inviting all the notables he could corral at a moment's notice.

Mademoiselle Marie, I noticed, was one of the guests and - as the Director-General introduced me around the festive board and explained to all and sundry how he was going to enjoy an automobile trip to Bucharest, instead of having to depend on a stuffy, overcrowded train - my eyes surreptitiously glanced in her direction. She was at the moment in earnest conversation with two of the guests, Colonel G...and his wife. The latter, towards the end of the lunch, when demi-tasses and cigarettes enabled the company to mingle freely, pressed me to spend a week or more on her estates, situated in the Transylvanian foothills. I looked around for Miss Temperament, intending to ask her if the colonel and his wife were the friends she had previously spoken to me about, but she had mysteriously vanished! Not knowing exactly what to do under the circumstances, the invitation was tentatively accepted, Colonel G...promising to notify me at Bucharest headquarters of the exact date of my proposed visit.

Preceded by car and driver, we finally got on board the ferry-boat. The Director-General and I were at once invited on the bridge as the captain's guests.

Hardly had we proceeded a mile upstream when, emerging from the lower decks where she had somehow managed to secrete herself, Mademoiselle Maria, properly equipped in somber travelling suit - she had been a mass of white fluffiness and ribbons at lunch - appeared before our astonished eyes and, tripping up the stairs to the bridge, curtsied mockingly and said: "Me voici! here I am! Mother gave me permission to go with you!"

Tableau!!!

A stranger in a strange land, there was nothing for me to say or do; I turned to my companion for help! The dear old man looked for an instant as though he might be going to have a stroke of apoplexy - his face and hands became purplish, veins stood out like whipcords along his neck; he sputtered, he fumed, he remonstrated - all in vain! The eccentric damsel eyed him quizzically, occasionally bestowing a roguish smile on me; not a single word more did she utter but her facial expression reminded me, in no uncertain manner, of her last parting shot to me that very same morning. "C'est la femme qui a le dernier mot" - it is woman who has the last word!

"Isn't it jolly?" her eyes were saying to me, "there's nothing you can do now except...what I want!" Having apparently fully exhausted his polite vocabulary, and in evident despair over the extraordinary situation - the intensity and forcefulness of his arguments and expostulations ruled him out at once as a possible accomplice in the plot against my....worn-out tires - the poor gentleman finally shrugged his shoulders in a gesture of utter helplessness and said: "C'est une enfant gatée, mon Colonel; il n'y a qu'a s'incliner!" - she's a spoilt child, Colonel; we can only give in to her!"

That apparently settled the matter. Descending to the main deck, I notified my driver that he would have to make room in the car for one more passenger. This was perhaps the only time I really blessed the military regulation which forbids an unsollicited answer from an enlisted man! My poor, efficient, and loyal driver merely said, "very good, sir!" - but he must have had murder in his heart; he has to repair blow-outs!

Never! never again was I going to discourse to incredulous European mamas on the freedom and liberty of our American girls! Here indeed was emancipation run amuck! The young Roumanian girl was literally throwing herself on my mercy and that of the navigation official; she was on our hands - we could not send her back because the ferry-boat docked all night at Braila, and her mother and cousin would already have left Galatz for Bucharest before the erratic daughter could return to the former city; we could not leave her alone and unprotected in Braila.....though goodness knows I felt in dire need of protection myself!....we had to do just what this wilful, spoilt, semi-old, semi-infantile girl had all along planned to have me do, namely, take her in my staff car to Bucharest!

The cobblestone sidewalks of Buzeu.

We accepted the inevitable! Loaded to the hilt and creaking under the combined weight of four passengers, including my driver, and several hundred pounds of luggage, the auto proceeded gingerly out of Braila and started it's two hundred mile run to the capital of Roumania!

Taking the wheel myself, with the driver beside me, gave the two Roumanians a further chance to argue the matter out - the Director-General took full advantage of the situation and lectured fore and aft, up and down - but verbal pyrotechnics did not lighten the car!

As night came upon us with the rapidity peculiar to the East, Mademoiselle Maria began to shiver with the cold - for the nights are quite cold in Roumania, even in summer time, and the car was an open one, minus side-curtains. She was urged to wear my army coat; then, later, that of my driver also - for she still complained. The official was well protected with warm overcoat and lap-robe; we, in front, shivered and drove the car, turn-and-turn about. Our headlights were very dim, the battery being badly in need of re-charging; the roads were indifferently bad, and the strain on our eyes made it necessary to steer by turns. Just before dark set in, we entered the small town of Buzeu and ate a hasty supper served on a small table set out on the cobblestone sidewalk in front of what seemed to be the best restaurant in that primitive part of the country. While we three sat, my driver munched a mixture of stale bread and "gold-fish" - army name for canned salmon - while walking around the car and closely examining it's set of tires! I noticed him mumbling to himself and wondered what prayers - or profanity - he was using as a means of persuading the worn-out casings and inner tubes not to loose any of their compressed air!

Of course, we had the inevitable blow-out!

Of course, we had the inevitable blow-out; my driver had evidently resorted to profanity during his exhortations! Destroyed bridges necessitated making a number of detours, and once we even got on the wrong road, in spite of our two nątive guides, but, eventually we entered the outskirts of the capital in the wee hours of the morning, passed the famous race-track on the Chaussée and drove along the Callea Vittoria and Bulevard Elizabetta to the center of the city. Our first thought was to deposit sleepy and thoroughly tired-out Mademoiselle Maria on her own doorstep - after ringing the bell and seeing the servants ushering her in; our second was to find a garage and beds for our weary bones to lie on! The Roumanian official directed us to the - then-dilapidated Hotel Majestic; there was no garage, but a night-watchman was installed in the car after we had removed all of our luggage. Here we parted company, the dear old man apologizing profusely for the "spoilt child" who had wished herself on me.

Two beds in one room was all we could get. It did not take long to settle down, and the difference in army status did not in the least disturb my slumber or that of my faithful driver!

In Bucharest, official duties took up all of my time for a week, and I had just begun a round of formal visits, while waiting to hear from the Galatz friends regarding the exact date of the proposed visit to their country estates, when the tragi-comical note...cancelling everything and dismissing me from her thoughts - forever and forever - was handed to me!

That the intense young partisan scorned all partisans of the Centrophile ex-Premier Marghiloman, had been made quite evident by her impassioned tirades against them when we discussed everything from Armistice to Zuika on board steamer, but....that she should carry her personal feelings to the extent of punishing me - a friendly stranger - for making a social call on erstwhile travelling companions with whom I had lunched in the city of Costanza, was quite, quite extraordinary!...

The house party fell through; Miss Temperament and I never met again!

"Donna e Mobile!"

His Excellency, the President

February 4, 1919

"You remember," said Doc, smiling reminiscently as he gave his pipe a few vigorous pulls in order to assure proper combustion of masculine nectar, "that one of my post-war missions was to assemble and convoy a field hospital or "Auto-Chir" from the Italian city of Bologna to Bucharest, capital of Roumania?

His Excellency, the President.

Does anybody know of a better way for three or four friends, spending an evening together - when wives and children have departed for summer resorts and husbands try to convince themselves that they are enjoying the peace and solitude of a "closed for the summer" town house - than to start, in military vernacular, "swapping lies"?

Given a background of approximately equal experience and knowledge of the world, with adventures shared in common and the instinctive desire which all red-blooded men possess - and try to keep a secret - for rough life without any of the, to them, useless and time-consuming frills which our better halves cling to as barnacles to a ship - this is not quite a correct comparison either, for barnacles are periodically scraped off while love of frills lies eternal in feminine mind - given such conditions I repeat, what is jollier than to purchase the wherewithal, cook one's own supper, solid, sustaining and simple, and then, coatless and collarless with sleeves rolled up to provide ample ventilation, sit around porch or library, pipe and mouth, and utter thoughts without censoring them?

Echo answers, what!

"Remember the bully-beef and spuds we used to warm up on our Primus lamp and eat while on our way to the Balkans?"

"Bet I do!" replied big George after gulping down a generous mouthful of fresh bass which Tom had that morning received from the northern lakes, his special contribution to our delightfully impromptu repast, "bet I do; wish I had a case of those army cans at home this minute, save shopping you know - beastly nuisance, shopping; worst part of bachelor's life."

"Doc, how do you think the Jugo-Slavs are going to end up with their united kingdom of Serbs, Croats, and Slovenes? Are they going to stick together or fall apart, like a poorly glued doll-house? You were there long enough to form an opinion; let's have it."

"It all depends, Tom." Doc has the sometimes exasperating habit of weighing his answers, even during the most informal exchange of opinions. "It all depends on where the levelling occurs; we never had a similar problem to solve at home, for our colonial ancestors came from a common source representing an approximately equal stage of civilization and experience in self-government. We started on the level - the same level I mean - and our United States was the natural result of common ideas and ideals. With Jugo-Slavia it is vastly different; Croatia and Slovenia have had the benefit - not entirely unmixed it is true - of Austro-Hungarian laws and institutions. Old Serbia, the dominant military force in the new combination, is still woefully behind the times. The Serbians, the masses, are a primitive people with all the virtues and faults which are the common heritage of the illiterate and superstitious! If....and it is yet too soon to answer the query, if the levelling is to be an upward one, from Serbia to Slovenia to Croatia, then well and good. If the contrary is to take place, I'm not so confident of the future. It might have been best to form separate kingdoms or principalities, leaving Serbia to solve her own problems in her own way, independently. After all, the mere size of a country has little or no bearing on the happiness and prosperity of it's inhabitants - look at Switzerland, Andorra, San-Marino, Monaco - you can walk around either of the last two independant countries in a few hours, yet you find contentment and a sturdy pride of fatherland and of native institutions almost incomprehensible to us Americans with our huge horizon! If confederation means the sacrifice of too many traditions, customs, and philosophies, I for one am in favor of smaller countries. The great charm of southeastern Europe lies in it's diversity; there's a certain elusive charm in people who are different from those who daily surround us. One gets tired of meeting the same cut of palm-beach suit in Maine and Florida and California, summer after summer - perhaps that's why so many of us trip it to Europe whenever we can, to forget the drab monotony of our cities and citizens. We are too standardized!"

"Pretty hot night, Doc," murmured big George with mock anxiety in his voice, "feeling all right?"

"Cool as a cucumber, thanks!....Take, for instance, those sturdy republicans living in the diminutive country of San-Marino, the smallest republic in the world, twenty-two miles square to be precise. For centuries they have lived on the rocky mountain overlooking the Adriatic sea, surrounded by Italian territory yet preserving their independance, customs, and, last but not least, costumes! They love their past, keep their castle and battlements intact, and live to-day much as their forefathers lived centuries ago - why should they become Italians? I can easily visualize what would happen to their quaint fortified town, were the San-Marinese - as the inhabitants are called - to subject themselves to Italian laws; the fascinating old walls and battlements levelled to make an outer drive around the town, telegraph and telephone poles and wires strung all over the mediaeval narrow streets, a cog-wheel railroad from the city of Rimini landing idle, gaping tourists by the thousand at their very doorsteps, hideous modern hotels ruining the picturesque landscape and nightly drowning the love-calls of countless frogs in the Marecchia valey below with their jazz-orchestra syncopations - modern clothing and plumbing, disinfected and pasteurized and chlorinated atmosphere everywhere..ugh!"

"Get a wet towel, Tom, Doc's engine is overheated!"

"Overheated nothing! As I was going to say when you fellows tried to interrupt me, now, thank God, the trip to San-Marino is a tedious, even somewhat dangerous one; autos have to remain outside the city gates and the few tourists who do go there, go to get local atmosphere, not remove it; long may it linger! Did I ever tell you how I really happened to visit the tiny republic - way back in 1919, after the Armistice? No? Well, fill up your pipes, here goes!"

Big George had, as a matter of fact, been to San-Marino himself, independently; Tom, poor fellow, although he tried hard, was unable to enter the service and had to be content with home duties while the rest of us were "over there", doing our various allotted bits for the common good of the great cause. It may therefore have been a mere transient touch of envy - or regret - which prompted him to interrupt the narrative.

"Never mind San-Marino, Doc," said he. "It will keep for another time; I asked you for your opinion on Jugo-Slavia - come back to where you were!"

"But this story illustrates my point!"

"All right! all right! fire away." And, Tom relapsed into an attitude of attentive silence.

"You remember," said Doc, smiling reminiscently as he gave his pipe a few vigorous pulls in order to assure proper combustion of masculine nectar, "that one of my post-war missions was to assemble and convoy a field hospital or "Auto-Chir" from the Italian city of Bologna to Bucharest, capital of Roumania?

I took command of the outfit before my hospital-corps men were assembled - as a matter of fact, big George and I were the sole members of the unit for several weeks. An excellent staff-car formed part of the rolling stock....but I had no driver! As preparations for the long trip to the Balkans necessitated frequent journeyings to the town of Padua, Mantua, and Vicenza, where the U.S. Army contingent in Italy kept it's supply depots, a request for an Italian military driver was sent in and duly honored. Efficient and amiable Valerio Sciutti became my driver and, both figuratively and literally...drove me to visit San-Marino!"

"Sounds funny," murmured Tom; "didn't you want to go?"

"Never had thought of the blooming place" retorted Doc. "I knew, vaguely, that there was some such rock and village on the west coast of the Adriatic, a hundred odd miles south-east of Bologna but...go there, no! I was impatiently waiting for my men to assemble and get started. Valerio persistently, day by day, in every way...O bother Coué; he's spoiling my English! Valerio injected dose after dose of curiosity-serum into me until I succumbed! No matter where we drove, no matter how much I might glory in and expatiate over the scenic, historic, or architectural beauties of northern Italy, the rascal would invariably shrug his shoulders contemptuously and throw a damper on my enthousiasm by saying: "wait until you have seen San-Marino, signor Colonello," putting his fingers to his lips and blowing a tender kiss in the air - presumably in the general direction of the aforementioned village!

It became a veritable obsession with him; see San-Marino or die! For a long time I could not understand why his thoughts should so constantly revert to this one spot on earth - then, one day, after a particularly long and tedious inventory with the final "taking over" formalities which made me the responsible officer for the entire field hospital - as we were driving back to Bologna from the aviation field where my trucks were temporarily stored under hangars - it occurred to me to ask Valerio what town he originally came from. Lucky thought! The youthful scamp grinned from one large ear to the other, blushed like a damsel and replied: "I am from Verucchio, signor Colonello; it is only five miles from San-Marino!"

"Capisco adesso! Now I understand! How very stupid of me not to have thought of this before. Call for me at eight tomorrow morning, prepared for a trip to your famous republic." It was quite a relief to know at last why San-Marino had been so persistently dangled before my mind's eye! Verucchio was on our way, we could not avoid passing through it.

Depositing me at my hotel door, the overjoyed driver hurried off to prepare a car for the journey. In a few minutes the pest was back again - looking more foolish and embarrassed than ever!

"Signor Colonello" said he, "I hardly dare ask the favor of you; you may think me too bold but - would you honor my father and mother by lunching at our house? We are so near San-Marino, and the restaurants there are so poor! May I send off a telegram to my father, telling him you will come? We have chicken and ham and sausage and fruit from our farm; our wines are excellent - vini spumanti, sparkling - ah! Signor Colonello, what pleasure you will give to my mother and to my sisters! Will you come?"

"Per Bacco, yes, yes! Go, presto, and notify your father!" Happy Valerio oozed happiness from every pore of his radiant face; had he not been in uniform

I would have expected that spontaneous osculation on both cheeks which, for some unfathomable reason, seems so natural amongst Latins, so uncouth with us!"

"Grazia! Grazia! Signor Colonello, thanks, for me, for my father, for my mother, for Catarina for.....

"Via, via! go, go! before I change my mind!" A rigid salute, at attention, and he had vanished!

A glorious sun shone on us as we left Bologna for our hundred mile run along via-Emilia, that wonderful modernized Roman road which, almost in a straight line, crosses northern Italy from Milan in Lombardy to Rimini on the Adriatic. There is all too much rain and mud in winter and spring-time in that part of the world; the fact that we were so exceptionally favored on that memorable occasion is worth recording. Our road took us through Imola, Castel-Bolognese, Faenza, Forli, and Cesena; all towns possessing considerable local color and individuality.

"You see," said Doc, pausing in his narrative to refill his pipe and swallow a luscious concoction of lemonade and ...No! I can't give you his name, he makes it for his own consumption and slips a gallon or two into my house in the wee hours of the night..."you see, individuality is what I love, in people as well as in towns. Each town in Italy is a distinct picture in itself - unique, mysterious, fascinating!"

The outskirts of Faenza had been turned into a British base, and there were hundreds of Tommies coming and going, giving the otherwise sleepy burg an unusually bustling aspect. How many hitherto unknown towns and villages have been made by the war, only to lapse again into peaceful slumber as soon as camp activities had disappeared! Pardon the interpolation, but the story of the outspoken proprietress of a small hotel at Paris-Plage, an unimportant bathing resort near Boulogne, is typical of these sudden war booms. When a certain British officer attempted to condole with the dame over the duration of the war and it's effect on France, she replied: "Je voudrais que la guerre dure encore une année; je pourrais alors me retirer et vivre de mes rentes - I wish the war could last another year; by that time I would be ready to retire and live on my income!" This remark was made in 1917; a year later it was currently reported that she had disposed of her modest holdings for the tidy sum of two million francs!

Beyond Faenza we motored through acre after acre of asparagus beds, a picturesque sight to gaze upon if you are not hungry at the time. Served with white sauce they are known all over Italy as Parma asparagus - Parma being undoubtedly selected as sponsor to the delicacy because it was two hundred miles away and was already famous for it's double violets! What's in a name!

Leaving the via-Emilia shortly before arriving at Rimini, Valerio turned our car to the right and drove up the charming little valley of the Marecchia which flows at the foot of Monte-Titano, on the Summit of which is perched the old fortified town of San-Marino. With a delightful grin on his swarthy face: "Ecco il mio paese!" said my driver, pointing to the small village on our left and half way up the mountain.

"Mio paese" - we have no expression which conveys the same tender meaning. My country, is too comprehensive, too sweeping a term; mio paese means the section of land in which you were born and in which your people are living or have lived. My home town, approaches the Italian expression, but what term can one use if one happens to have been born on a farm?

Verucchio....consisted of a Piazza

The Malatesta Castle or "Rocca", overlooking Rimini.

Verucchio, like every other mountain village in Italy, consisted of Piazza or square around which were clustered the few stores in which supplied the villagers wants, a Municipio or city hall, a church, the inevitable public fountain in the center of the piazza and a half dozen narrow streets darting up the side of the mountain in a variety of grades and fantastic curves. The one and two story houses on either side had adapted themselves as best they could to the inequalities of the rocky soil on which they were built. If, as occasionally occured, there didn't happen to be enough room on one side for a complete habitation, the street itself was arched over and you drove or walked under signora's bedroom or kitchen! The daily mail and stage coach combined jingled up from Rimini as we drove through the quaint piazza; Valerio, fully sensing his immense importance and with the love of the dramatic which is an inherent part of Italian temperament, shouted a joyous "buon-giorno!" - good-day - to the assembled villagers and sent the car spinning up one of the side streets, hugging remains of old battlements, climbing higher and higher up until, opposite a modest but very clean two-story house, he came to a sudden halt.

Jumping from the car, he opened the front door and shouted: "Ecco il tuo Valerio! behold thy Valerio!" Down the steps came a buxom housewife, three very handsome daughters, an equal number of sons, and the father - the latter a tall, thin, intellectual-looking man who might have readily passed off for a Connecticut Yankee, were it not for his black eyes and fierce mustache.

Guiseppe Sciutti was quite evidently not of the peasant class; neither was he a signor - a few diplomatic questions soon enabled us to classify him in the Italian social scale. In addition to being a small landowner, he acted as overseer or agent for several absentee landlords who owned large sections of Verucchio commune and spent their incomes in Rimini.

We were received with simple, dignified courtesy and immediately ushered into the combination parlor and dining room where the festive board had already been laid. While mother and daughters disappeared to finish the preparation of the hot dishes, our host filled the Vermouth glasses and toasted all America - from the President to ye humble narrator! The necessary counter-toast to Italy, the King, Verruchio and the Sciutti family having been drunk, we attacked the "anti-pasti" or hors d'oeuvres of which there were many. With genuine and very pardonable pride, the elder Sciutti informed us that everything we were eating and drinking came from his own farm. Restrictions were still enforced in all Italian cities four months after the armistice, but the outlying villages were not suffering - far from it.

With a heaping platter of fried chicken and delicious vegetables passed and repassed around the table by three lovely maidens, it was hard to be modest! This remark may savor of plain gluttony; before passing sentence however, kindly bear in mind that some of us had lived on the war diet of the Allies in 1917. We ate and drank and ate again for two blessed hours! Guiseppe Sciutti felt called upon to apologise for what was, to him, an unpardonable omission! After the last mouthful of "dolci" or dessert had been swallowed and I had beseeched Catarina - who, as the oldest daughter had centered her ministrations on me - not to temp me with any more cake, the dear old man implored my forgiveness and begged me to excuse the..absence of..toothpicks! "I assure you, signor Colonello" said he with utter despair in his voice "there was not a toothpick to be found in all Verucchio! Oh! this war has been terrible!"

Apparently as the result of some prearranged signal, the Sindaco or Mayor of the village appeared upon the scene with the coffee and cognac. Like many other Mayors in Italy - who have to be property owners though not necessarily residents, he lived in Rimini, driving up to Verucchio at stated intervals in order to preside at council meetings. In honor of America, he had opened his house for us across the street and, after quaffing Sciutti's cognac, he escorted the entire party to his own dining room where we had to repeat the same performance and drink additional toasts!

Just as I was about to give the signal to start for San-Marino, a delegation of leading citizens appeared to chaperone us through the ruins of the Malatesta castle or "rocca di Malatesta". The half-mile walk down to the piazza and then up a winding mule path to the tower, dungeon, and council-chamber, the latter now used as a village theater, fully restored our equilibrium - Italian cognac is more fiery than it's French prototype!

The view was certainly worth the effort. From the tower one absolutely dominated the city of Rimini. This delightful sleepy little village of Verucchio was once the chief stronghold of the Malatesta family, lords of Rimini from the twelfth to the sixteenth centuries. Malatesta may mean nothing to the average layman, but who has not heard of Francesca da Rimini, so pathetically immortalized by Dante in his Inferno? You doubtless remember how she, a beautiful, proud, and high-spirited girl, was tricked by her own father into marrying - for political reasons - the deformed, repulsive, and degenerate Giovanni Malatesta, count of Verucchio and one of the sons of the then ruling despot of Rimini. Giovanni's brother Paolo, handsome as Apollo, was sent to Ravenna where he impersonated his older brother during the wedding ceremony. Only on the morning after the wedding and return journey to Rimini did the unfortunate Francesca discover the dastardly trick which had been played on her. Not long afterward, the cripple Giovanni discovered an intrigue which had grown up between his wife and brother Paolo and killed both with the same dagger!

With such wealth of tradition surrounding them on all sides, is there any wonder that one encounters the poise of experienced men and women of the world amongst the simplest villagers of old Europe? Poise seems to be the best word available. The Verucchians made us feel that we were conferring an honor upon them by deigning to spend part of one day in their midst, but they showed no humility and evidenced a readily understood pride in the historic associations which had been their birthright. Our corresponding classes lack this poise, it comes only with tradition and the latter cannot be machine-made!

A hurried visit to the war-orphans, a last glass of Vermouth in the club-room of the old Garibaldi veterans - newly ornamented with photographs of natives who had fallen in the World War - and off we finally went, winding around hills, crossing gullies, up, up through the lower village of Borgo, to the top of Monte-Titano, 2635 feet above sea level. What a panorama! On three sides the Apennine range of mountains; in front of us Rimini and the Adriatic. At sunrise, on a clear day, the Dalmatian coast is said to be plainly visible.

What a panorama!

Emperor Diocletian was a Dalmatian; after tolerating Christianity for a few years he finally ordered a general massacre, about the year 303 A.D. Tradition hath it that a converted stone-mason of Rimini, Marinus by name, objected to being fed to the lions and wandered up to the summit of Monte-Titano where he hid in a cave and became a hermit. As it was then the fashion for hermits to acquire supernatural powers and become saints, he did likewise, performed miracles, founded a monastery, died and was, in due course of time, canonized!

Thus originated San-Marino, oldest and smallest republic in the world. If you don't believe it "look, look in the book and see" - as the old song sayeth! It is unquestionably a genuine antique. It's three towers still seem to defy the world at it's feet; the old castle is of pure Ghibelline architecture and only the Palazzo del Governo - or government house - is a clever imitation of a medieval construction. The revenue derived from the sale of fruit, wine, and silk-worms has long been surpassed by that extracted from tourists! Stamps, coins, flags, coats-of-arms, and picture-post-cards keep the little community well supplied with funds for it's modest exchequer. The Italian government allows a fixed sum per annum for the tobacco monopoly, but does not interfere with the internal revenues of it's peaceful wart - what else can one call this excrescence, surrounded as it is on all sides by Italian territory? San-Marino used to be a favorite asylum for criminals and political refugees, until Italy compelled it to expell all strangers after a three days' sojourn, unless their identification papers were satisfactory. The great Garibaldi himself, in the days when he was looked upon by the authorities of un-united Italy as a soldier of fortune, before becoming the national hero of united Italy, tried to hide in San Marino but was sternly ordered to "move on" by the local authorities. Just what dire threats were used in order to obtain this law was impossible to ascertain; shutting off the San Marinese tobacco allowance sounds probable!

In front of the Palazzo...stands a monument to Liberty.

Should any discontented American snob desire to add a handle to his - or her - name, San Marino will graciously oblige...for a consideration! The "Reggenti" or presidents have the power of conferring sonorous titles upon all and sundry persons who might express a craving for this weakness. In front of the Palazzo del Governo stands a small monument to Liberty. By the simple process of donating this block of quite indifferent sculpture, plain Mrs S-W...please note that she was already hyphenated, became Duchessa di Aquaviva....a fresh-water duchess, so to speak!

While contemplating this silent reminder of a universal craving to be "different from others", a gorgeous flunkey approached us, cockaded hat in hand, and asked if we did not wish to meet "sua Eccelenza il Presidentc". Receiving an affirmative reply, he bowed himself off while we continued with our sight-seeing. The unique location, mountain scenery, old battlements, and towers constitute the main attractions of San-Marino; it's church was a disappointment. As we were hurrying out of it, our self-important flunkey returned with the announcement that his excellency would be pleased to see us at once. Back we went to the government house where we were ushered into the large and handsomely decorated atrium and told to wait. Pretty soon a door opened onto the inner balcony above us and a second, equally well costumed flunkey descended the grand stairway with dignified precision to announce that his Excellency.....the Minister of Foreign Affairs was awaiting us!

Atrium of the Government Palace with grand stairway.

Yes, laugh all you want; the flunkey said the Minister of foreign affairs; isn't San Marino an independent country?

We dared not look at each other; this was too solemn an occasion - biting our lips to control our facial muscles, we marched upstairs!

At the entrance of his private office stood a dapper little man wearing a frock coat and brown cotton gloves - of course he had other bits of clothing hanging about his person, but they didn't impress one to the same extent! He clicked his unspurred heels together; we did likewise with our spurred ones and then "Sua Eccelenza il Ministro delle affare Esteri" - to give him his full title in Italian - asked whom he should have the pleasure of announcing to his excellency the President. Upon my reeling off rank and names of our party, he gave us a courtly bow, backed off across his office, and disappeared from view through a rear door, appearing a few minutes later with the announcement that his Excellency the President was ready to receive and welcome us in the name of San-Marino.

With bated breath and firmly compressed lips we marched into the president's sanctum, a small but artistically furnished room having a raised platform at it's further end. A very official-looking desk and high-backed chair occupied one side of this sort of dais and, standing besides the furniture was "Sua Eccelenza", a highly intelligent-looking man of about fifty years of age, wearing the same cut of frock-coat and the same brown cotton gloves.

The two Presidents or "Cappitani Reggenti".

Had our visit not been such an impromptu one, the president would undoubtedly have donned his official costume and summoned his colleague - for San Marino, instead of electing a vice-president, chooses two presidents who are called the "Cappitani Reggenti". It is possible however that, it being summer-time, all of the precious mediaeval costumes had been carefully wrapped up and put away with dried lavender blossoms instead of our nauseating moth-balls!

Behind his excellency, on the wall - still dangling to-and-fro as the result of it's all-too-evident hasty resussitation from some hidden cupboard - hung the photograph of our president! Our lips blanched from the accentuated compression which suddenly became necessary!

The minister of foreign affairs evidently drew double pay, for he also acted as official introducer. After the formal presentations were over, the President began a stereotyped eulogism of our country's many virtues and charities - right hand stuck in frock coat, a la Napoleon crossing the Alps - and then, after we had thanked him for his kind words and were making ready to depart, the chief executive turned towards his right bower and said "the minister of foreign affairs will now have the floor" - or words to that effect!

More formal bows and then, off went the diplomat of San-Marino! He started off like an athlete in a hundred yard dash but kept on running, verbally, long after he had broken the tape line! He super-eulogised America, showered thanks and benedictions on us, individually and collectively, and then - which was really interresting - recounted the glorious part which the volunteers of his diminutive country had played in the war. Twenty-five young San-Marinese had joined the Italian army; four of them had been killed and several wounded; the Minister himself had only recently received an honorable discharge from the same army. The final bit of information, to the effect that San-Marino had really never declared war on the Austro-Hungarian empire - as our papers at home had reported to be the case - proved that, after all, there was some sense of humor among the elite of this sturdy little republic!

Again thanking the president and his minister for according us this most interresting and unique interview, we signed the visitors book and departed. After all, why shouldn't they take themselves seriously? San-Marino has it's own standing army and guard of honor, both wearing mediaeval uniforms. Not so many generations back, all of Italy was subdivided into similar republics, "Villafrancas" or free cities, dukedoms, and so forth. Italy has changed, San-Marino has not. If some eight thousand odd republicans choose to remain independent and maintain a minister of foreign affairs - why it's nobody's business but their own!

"Very fine Doc! Very fine!" grumbled Tom, "but what has your blooming republic of San-Marino to do with the future of Jugo-Slavia?"

"For one thing, it's only across the Adriatic! For another....I was discussing individuality. Aren't we speaking our thoughts to-night? Yes! well then, don't thoughts ramble? Jugo-Slavia will keep cool...not so the lemonade..."

Pass the Jug!

Letter Monsieur le Colonel en Rezerve

In response to 8 Janvier, 1920 - Roumanian

January 8, 1920

LE GRAND QUARTIER ROYAL
Bureau de Décorations

No 5830
1920 Luna Martie zlua 3
MARELE CUARTIER REGAL
Biuroui Decoratillor
către

Monsiuer le Colonel en rézerve
GEORGE DE TARNOWSKY

Service de Santé Americaine
Nr. 30 North Michigan Avenue
Chicago

En réponse à votre letter du 8 Janvier 1920, au nom de SA MAJESTE LE ROI, je vous adresse les meilleurs remerciements et je suis heureux de vous envoyer les insignes de l'Ordre "l'Etoile de Roumanie", au grade d'officier avec glaives, en y joingnant mes chaleureuses félicitations.-

Veuillez agréez, Mon Colonel, l'assurance de ma consideration la plus distinguée.-

PAR ORDRE DE SA MAJESTE LE ROI
Chef du Bureau des Décorations
Aide-de-camp de S.M. le Roi
COLONEL

War Department Letter

Roumanian King – Chief of Staff Military Division

April 2, 1920

MILITARY INTELLIGENCE DIVISION

In replying refer to
2447-V-6
M.I.5-2

WAR DEPARTMENT
OFFICE OF THE CHIEF OF STAFF
WASHINGTON

April 2, 1920.

Mr. George Detanouwski,
30 North Michigan Avenue,
Chicago, Illinois

My dear Mr. Detanouwski:

The American Military Attache at Bucharest, Roumania, has forwarded to this office a decoration which the King of Roumania has conferred upon you in recognition of the extraordinary services performed by you during the war, while serving with the Red Cross attached to the Roumanian Army. This Division takes pleasure in forwarding the decoration to you to-day by separate registered mail. Enclosed herewith is a letter address to you by the Roumanian Government.

It is requested that you acknowledge receipt of this decoration to this office.

Very truly yours,
M. Churchill
Brigadier General, General Staff
Director of Military Intelligence
Assistant Chief of Staff

By:
Sherman Miles
Colonel, General Staff

1 encl.
iav

Roumanian Decoration

April 9, 1920

April 9th '20

From: George de Tarnowsky Col. M.C. Reserve Section
To: Military Intelligence Division Washington D.C.
Subject: Roumanian Decoration

2447-V&6
M.I.5-2

1. Receipt is hereby acknowledged of letter from the Roumanian government and of decoration from same source.

George de Tarnowsky
Col. M.C. Reserve Section U.S.A.

Genealogie de la Famille Tarnowski

French Version

Overview

GENEALOGIE DE LA FAMILLE TARNOWSKI DE TARNOW EXTRAIT DE L'ALMANAQUE LE GOTHA.

Famille catholique en Autriche (Galicie) et en Russie polonaise (Volhynie). Celebre famille polonaise. Jean Amor Magnus de Tarnow, nommecomte de l'Empire par l'empereur Charles V (decre donne a Augsbourg le 24 Decembre 1547) Grand Hetman des armees polonaises. Jean Hyacinthe Comte Tarnowski, nomme par l'empereur Joseph 11 (decre donne a Vienne le 18 Juillet 1786) Staresten de Bychow et Hahorlit. Blason: d'ajur, avec un cornet retourne vers le sommet; croissant d'or au dessus duquel brille une etoile egalement d'or; couronne de comte; le casque couronne est surmonte d'un buisson de plumes de paon; devise: "Ad astra tendite viri."Une des familles les plus renommees, puissantes et celebres de la Pologne, dont l'histoire se mele a celle de ce pays depuis le Xlll ieme siecle. Spithko de Tarnow (1290-1356) fut premier conseiller du roi Vladislas lV le Bref, castellan de Cracovie et fondateur de la ville de Tarnow; ilepousa Gertrude, nee Duchesse de Majovie et Siradin de la lignee royale des Piast, niece du roi Lech le Noir et Vladislas le Bref. Son oncle Jean Amor, avec le surnom de "Balidus", palatin de Cracovie (1390-1485) epousa Elizabeth,nee Comtesse Sternberg, fille du Grand Chancellier de la Couronne de Boheme, Jaroslaw Comte de Sternberg, et soeur dela femme du roi de Boheme Georges Podiebrad. Elle porta en dot le duche de Randnitz en Boheme qui resta en possession des Comtes de Tarnow pendant 150 ans.Le petit-fils de Jean Amor, fut Jean Amor Magnus (1488-1561)premier conseiller de le couronne de Pologne sous le regne des deux derniers Jagellons, les rois Sigismond l et Sigismond ll. Auguste, grand general du royaume de Pologne, castellan de Cracovie, fondateur des villes de Tarnopol et Tarnograd, et due de Randnitz, fut cree comte de l'empire le24 Decembre 1547. Sa fille Sophie, qui epousa le Prince Constantin Ostrog Ostrowski, acheta tout le duche de Randnitz en 1574.

Michel, comte de Tarnow Tarnowsky (1590-1660) castellan Mojnies, epousa Anne, fille du Comte Emmorich Csobor de Szent Michaly, veuve de Stanis-las

Comte de Turzo, devient possesseur par sa femme des immenses proprietes de Csobor etSzent Michaly en Hongrie, qui jusqu'a l'annee 1750 demeurentdans la famille et qui furent achetes par le Comte Batthyany. L'empereur Ferdinand lll, par decret donne a Presbourg (24 Juin 1655) le nommat "Magnat de Hongrie" et Comte de Gross Csobor, titre qui fue ensuite confirme a Vienne le 25 Juillet 1804 en faveur du Comte Jean Felix.Ses deux petit-fils furent les fondateurs des deux branches encore existantes.Le Comte Michel Hyacinthe Amor (ne 1675 mort 1718) fonda la branche ainee qui se fixa en Galicie.Le Comte Alexandre Dominique Amor (ne 1676 mort 1710) fonda la branche cadette qui se fixa dans la Pologne russe.Wiclowies, depuis 1305 le fief de la famille et Tarnobzjeg avec le chateau Djikow recu en partage en 1522, restaurant ala branche ainee.Comte Gaetan Amor de Tarnow, epousa Anastasie de Bobusj; fils: Comte Antoine Fortune, epouse Suzanne nee Comtesse Jekewieka; petit-fils Jean, mort 1835, epousa Pauline de Wistocka; arriere petit-fils Jean Theodore, epousa Isabelle nee Comtesse Prazmow-Prazmowska.

Genealogy of the Tarnowski Family

English Translation

Overview

As translated by Nixon de Tarnowsky,
son of Dr. George de Tarnowsky

GENEALOGY OF THE TARNOWSKI FAMILY OF TARNOW EXTRACT FROM THE ALMANAQUE LE GOTHA

Catholic family in Austria (Galicia) and in Polish Russia (Volhynia). Famous Polish family.

John Amor Magnus of Tarnow, appointed Count of the Empire by Emperor Charles V (decree given at Augsburg on December 24, 1547) Grand Hetman of the Polish armies.

Jean Hyacinthe Comte Tarnowski, appointed by Emperor Joseph 11 (decree given in Vienna on July 18, 1786) Staresten of Bychow and Hahorlit. Coat of arms: d'ajur, with a cornet returning to the top; golden crescent above which shines a star also of gold; count's crown; the crown helmet is surmounted by a bush of peacock feathers; motto: "Ad astra tendite viri."

One of the most renowned, powerful, and famous families in Poland, whose history has been intertwined with that of this country since the 13th century. Spithko of Tarnow (1290-1356) was first adviser to King Vladislas lV the Short, castellan of Krakow and founder of the city of Tarnow; he married Gertrude, nee Duchess of Majovie and Siradin of the royal line of Piast, niece of King Lech the Black and Vladislas the Short. His uncle Jean Amor, nicknamed "Balidus", Palatine of Krakow (1390-1485) married Elizabeth, nee Countess Sternberg, daughter of the Grand Chancellor of the Crown of Bohemia, Jaroslaw Count of Sternberg, and sister of the wife of the King of Bohemian Georges Podiebrad. She bore the Duchy of Randnitz in Bohemia as her dowry, which remained in the possession of the Counts of Tarnow for 150 years. Two last Jagiellons, Kings Sigismund l and Sigismund ll. Auguste, great general of the kingdom of Poland, castellan of Krakow, founder of the cities of Tarnopol and Tarnograd, and due of Randnitz, was created count of the empire on December 24, 1547. His daughter

Sophie, who married Prince Constantin Ostrog Ostrowski, bought the entire Duchy of Randnitz in 1574.

Michel, Count of Tarnow Tarnowsky (1590-1660) castellan Mojnies, married Anne, daughter of Count Emmorich Csobor de Szent Michaly, widow of Stanislas Count de Turzo, becomes owner by his wife of the immense properties of Csobor and Szent Michaly in Hungary, which until the year 1750 remained in the family and which were purchased by Count Batthyany. The Emperor Ferdinand III, by decree gave Presburg (June 24, 1655) the name "Magnate of Hungary" and Count of Gross Csobor, a title which was then confirmed in Vienna on July 25, 1804 in favor of Count Jean Felix. -son were the founders of the two still existing branches. Count Michel Hyacinthe Amor (born 1675 died 1718) founded the elder branch which settled in Galicia. Count Alexandre Dominique Amor (born 1676 died 1710) founded the younger branch, which settled in Russian Poland. Wiclowies, since 1305 the stronghold of the family and Tarnobzjeg with the castle Djikow received in division in 1522, restoring to the elder branch.

Count Gaetan Amor de Tarnow, married Anastasie de Bobusj; son: Count Antoine Fortune, wife Suzanne, nee Countess Jekewieka; grandson Jean, died 1835, married Pauline of Wistocka; great-grandson Jean Theodore, married Isabelle, nee Countess Prazmow-Prazmowska.

The Six Hundredth Jubilee

May 31, 1930

The six-hundredth Jubilee of the Tarnowksi family seat.

The city of Tarnow in little Poland (called Galicia by the Austrians), age-old cradle of the Tarnowski family, is observing this year its six hundredth anniversary.

In the year 1330, Wladyslaw Llokietek, King of Poland, signed an act of erection on the strength of which the village of Tarnow-Wielki was elevated to the rank of City.

The first "Master of Tarnow" was the Palatine of Krakow, Spytko, or Spicimir, of the Leliwit lineage. To these ancient times dates back also the castle on St. Martin's hill, erected and occupied, as the documents say, by the first rule of the city of Tarnow, in the year 1340.

The definite reign of the "Masters of Tarnow" was begun by the son of Spytko, Rafeal, who took over the city authority from the hands of the hereditary bailiff. The reign of the "Masters of Tarnow" lasted then for 423 years until the Austrian government put an end to it.

The third ruler of the city, Jasko of Tarnow, immortalized himself in the annals of our history by bringing about the marriage of Jagiello, Grand Duke of Lithuania, to Jadwiga, Queen of Poland. Still more renowned was the Commander-in-Chief of the King's armies who received the name Tarnowski from the Jagellonian kings, and upon whom the Emperor of the Holy Roman Empire, Charles V, bestowed the title of Count. From that time on, Tarnow became the principal abode of the Tarnowski family. The last absolute master of Tarnow was Heronimus Sanguszko who, in 1781, was forced to give over his role of "The highest authority" to the magistrate of the Austrian Government.

The 31st day of July is remembered in the annals of the city of Tarnow as the day in which the Austrian soldiers entered and proclaimed the Maria Teresa act of dismemberment of Poland.

From the year 1783, the city expanded considerably, becoming a Bishopric. In 1866, the city was granted autonomy. The power of the government Burgomaster came to an end, and the first local Burgomaster elected was Dr. Wojciech Bandrowski. The rule of these later masters of Tarnow extends to Michel Skowronski who holds office in this year of the 600th Jubilee of the founding of the city.

1930

Distinguished Service Medal – Recognition

Documents from Presentment of Distinguished Service Medal

August 5, 1921 and June 23, 1922

August 5,1921.

From: The Surgeon General.
To: The Adjutant General of the Army.
Subject: Recommendation for the Distinguished Service Medal.

1. During the operations of the American Expeditionary Forces in France a recommendation was submitted from the Headquarters of the S.O.S. at Tours for the award of a Distinguished Service Medal to

LIEUTENANT COLONEL GEORGE de TARNOWSKY, MEDICAL CORPS

for exceptionally meritorious and distinguished services to the Government of the United States in a position of great responsibility during the world war.

2. This case was considered along with many others during the very busy post-armistice period and failed to receive favorable action at that time.

3. Since my return from France I have had an opportunity to soberly reflect upon and make an analytical study of the problem that confronted the Medical Department overseas, with a view to determining as far as practicable just what features in our organization were responsible for the good results achieved. In summing up my findings I am of the opinion that the distinguished service of some officers of the Medical Department whose responsibilities and accomplishments overseas were of an outstanding character, have not yet been adequately recognized.

4. As the name of Colonel de Tarnowsky clearly falls within this category I urgently recommend that his case be reconsidered, with a view to awarding him a D.S.M. for his eminent services in directing American Red Cross Military Hospital 45, A.E.F., as cited in the recommendation which should now be on file in your office.

5. That the A.E.F. was able to achieve success in the care of battle casualties evacuated to the Paris district during June and July 1918, was to a very large extent due to the individual efforts and vision of this officer.

6. In civil life, to which he has returned, he is a distinguished physician.

N.W. Ireland,
Surgeon General.

DISTINGUISHED SERVICE MEDAL

GEORGE deTARNOWSKY, lieutenant colonel, Medical Corps, United States Army. For exceptionally meritorious and distinguished services in organizing and commanding American Red Cross Military Hospital No. 5, American Expeditionary Forces. Notwithstanding the serious deficiency in personnel and material, Col. deTarnowsky, by his energy and efficient administration, received and treated great numbers of wounded evacuated to the Paris district during June and July, 1918.

IN REPLY REFER TO:

WAR DEPARTMENT
Headquarters
CAMP CUSTER, MICHIGAN.

July 25th, 1922.

Lieut. Col. George deTarnowsky,
Officers' Reserve Corps,
Camp Custer, Michigan.

My dear Colonel: -

I have just seen the enclosed order and I hasten to congratulate you on the honor which has just come whereby you are to receive the D.S.M. I recall very well what a wonderful piece of work you did in organizing and commanding American Red Cross Military Hospital No. 5 and I am glad to see Uncle Sam properly recognize your services.

In a radiogram to Corps Area Headquarters I have suggested that the decoration be sent here so as to permit us to make the award in connection with the proper formation of this Command.

Very sincerely yours,

VHM/S

[G.O.27.]

GENERAL ORDERS, No. 27.}

WAR DEPARTMENT
WASHINGTON, *June 23, 1922.*
Section.

I_ _Award of distinguished-service cross. -By direction of the President, under the provisions of the act of Congress approved July 9, 1918 (Bul. No. 43, W. D., 1918) as amended by the act of Congress approved April 7, 1922 (Bul. No. 6, W. D., 1922), a distinguished-service cross was awarded by the War Department to the following named enlisted man:

Clay Eversole, 1543587, private, Medical Detachment, 148th Infantry, 37th Division. For extraordinary heroism in action near Cierges, France, September 27, 1918. During the advance of Company K, 148th Infantry, Pvt. Eversole frequently exposed himself to great danger by carrying a number of wounded men through heavy machine-gun fire from an open field to a place of safety. Residence at enlistment: 435 Gunckle Avenue, Dayton, Ohio.

The citation for gallantry in action published on page 5, of War Department General Orders, No. 19, 1921, relating to Pvt. *Clay Eversole* for gallantry in action near Cierges, France, Sept. 27, 1918, is rescinded.

[A. G. 201.]

II_ _Award of distinguished-service medal. -By direction of the President, under the provisions of the act of Congress approved July 9, 1918 (Bul. No. 43, W.D., 1918), a distinguished-service medal was awarded by the War Department to the following-named officer of the French Army:

M. L. Delaunay, captain, French Army. For exceptionally meritorious and distinguished services while serving in the capacity of liaison officer with the 78th Division, American Expeditionary Forces. Home address: France.

[A. G. 336.4]
111763*-22

[G.O.27]

III_ _Awards of distinguished-service medal. -By direction of the President, under the provisions of the act of Congress approved July 9, 1918 (Bul. No. 43, W. D., 1918), as amended by the act of Congress approved April 7, 1922 (Bul. No. 6, W. D., 1922), a distinguished-service medal was awarded by the War Department to the following-named officers:

Louis B. Wilson, colonel, Medical Corps, United States Army. For exceptionally meritorious and distinguished services as assistant to the Director of Laboratories and Infectious Diseases, American Expeditionary Forces. He organized most efficiently a pathological service throughout the American Expeditionary Forces in France that was of inestimable value to the medical and surgical services. Residence at appointment: 830 West College Street, Rochester, Minn.

Robert Low Bacon, lieutenant colonel, Cavalry, O. R. C., then major, Field Artillery, United States Army. For exceptionally meritorious and distinguished services. As assistant to the Chief of Field Artillery from February 8, 1918, to January 2, 1919, he planned, instituted, and supervised the system by which an adequate number of properly qualified officers were secured for the Field Artillery. Residence at appointment: Westbury, Long Island, N. Y.

George deTarnowsky, lieutenant colonel, Medical Corps, United States Army. For exceptionally meritorious and distinguished services in organizing and commanding American Red Cross Military Hospital No. 5, American Expeditionary Forces. Notwithstanding the serious deficiency in personnel and material, Col. *deTarnowsky*, by his energy and efficient administration, received and treated great numbers of wounded evacuated to the Paris district during June and July, 1918. Residence at appointment: 2719 Pine Grove Avenue, Chicago, Ill.

Wesley W. K. Hamilton, lieutenant colonel, Adjutant General's Department, then major and lieutenant colonel, Coast Artillery Corps, United States Army. For exceptionally meritorious and distinguished services as adjutant, base section No. 3, in 1917-18. In the initiation, development, extension, and administration of those headquarters, through whose agency vast supplies and over a million troops were forwarded to France, he rendered a service of great value. Address: Care of The Adjutant General of the Army, Washington, D. C. Appointed from the Army. Residence at appointment: Dayton, Ohio.

[G.O.27.]

Charles Elliott Warren, lieutenant colonel, Ordnance Department, United States Army. For exceptionally meritorious and distinguished services while in charge of finances of the Small Arms Division, Ordnance Office, where his eminent ability as a financier and as an executive of large affairs, and also his previous military training were invaluable in the early organization of the division. Later, as Chief of the Small Arms Section, Procurement Division, Ordnance Office, and as one of the members and vice governor, the War Credits Board, Office of the Secretary of War, he rendered conspicuous service in the conduct of its immense affairs. Residence at appointment: 60 East Forty-second Street, New York, N. Y.

Edgar B. Colladay, major, Adjutant General's Department, then lieutenant colonel, General Staff Corps, United States Army. For exceptionally meritorious and distinguished services. As executive officer of the Cable Section, General Staff, during the World War, his sound judgment, his marked ability, his untiring energy, and his willingness to accept grave responsibilities, contributed greatly to the successful operation of the cable business during the World War. Later, as executive officer of the Statistics Branch, General Staff, he assisted materially in collecting and compiling valuable statistics pertaining to the World War. Address: Care of The Adjutant General of the Army, Washington, D. C. Entered military service from Wisconsin.

Campbell B. Hodges, major, then colonel, General Staff Corps, United States Army. For exceptionally meritorious and distinguished services. As acting chief of staff, 31st Division, he demonstrated the highest professional attainments, and through his zeal and never-failing tact he was responsible in a large measure for the development of the high efficiency of that division. Later, as a member of the War Department General Staff he rendered most valuable service in the development of the present system of efficiency reporting and in the development of the system of classification of commissioned personnel now in use. Address: Care of The Adjutant General of the Army, Washington, D. C. Entered Military Academy from Louisiana.

Stanley Koch, major, Quartermaster Corps, then lieutenant colonel, Infantry, United States Army. For exceptionally meritorious and distinguished services as executive officer the Remount Service, American Expeditionary Forces. By his zeal, efficiency, and devotion to duty, he has rendered great and beneficial service in the

[G.O.27.]

organization of the Remount Service in the American Expeditionary Forces. Address: Care of The Adjutant General of the Army, Washington, D. C. Entered Military Academy from Montana.

William F. Vordi, major, Medical Corps, United States Army. For exceptionally meritorious and distinguished services as surgical consultant and specialist in surgery of the chest in hospital formations at the front during the operations on the Marne and the St. Mihiel and Meuse-Argonne offensives. Residence at appointment: 27 Elm Street, New Haven, Conn.

Franklin L. Whitley, major, then lieutenant colonel, Infantry, United States Army. For exceptionally meritorious and conspicuous services. In 1917, before instruction pamphlets were issued, Col. *Whitley* prepared combat drill formations, suitable for war strength companies armed with new weapons. As battalion commander during the operations near Chateau-Thierry, he rendered valuable service. Due to his initiative and personal leadership, 39 days of constant contact with the enemy failed to break the morale of his organization. After the armistice, as Chief of the Decorations Division, General Headquarters, American Expeditionary Forces, by his sound judgment, professional knowledge, and exceptional ability, this officer performed his manifold and responsible duties with the utmost efficiency. He has rendered services of material worth to the American Expeditionary Forces. Address: Care of The Adjutant General of the Army, Washington, D. C. Entered Military Academy from Missouri.

[A.G. 210.52, D.S]

IV_ _Citation for gallantry in action. -By direction of the President, under the provisions of the act of Congress approved July 9, 1918 (Bul. No. 43, W. D., 1918), the following-named enlisted man is cited for gallantry in action and a silver star may be placed upon the ribbon of the victory medal awarded to this enlisted man. The citation is as follows:

John E. Kamb, 67089, corporal, Company B, 103d Infantry, 26th Division. For gallantry in action near Chatillon-Sur-Lein, France, August 20, 1918. Corp. Kamb, while in charge of a squad of men, in conjunction with Corpl. Kierstead, voluntarily went out in front of a well-fortified enemy machine-gun nest and succeeded in capturing and killing the entire crew. Residence at enlistment: Cambridge, Mass.

[G.O.27.]

[A.G. 201.]

BY ORDER OF THE SECRETARY OF WAR:

JOHN J. PERSHING,
General of the Armies
Chief of Staff.

OFFICIAL:

ROBERT C. DAVIS,
Acting The Adjutant General.

WASHINGTON: GOVERNMENT PRINTING OFFICE: 1922

Letter from the Commander-in-Chief

April 20, 1919

American Expeditionary Forces
Office of the Commander-inChief

April 20th, 1919.

My dear Colonel:

I write to express my appreciation of the valuable services which you rendered the American Expeditionary Forces.

As Commanding Officer of the American Red Cross Military Hospital No. 5, you displayed marked ability for organization and administration, and the efficiency of that unit reflects the greatest credit upon you. At all times, you were constant in devotion to your exacting tasks and untiring in your energy. Your efficient service resulted in the alleviation of the suffering of hundreds of our sick and wounded soldiers.

I regret that I was not able to thank you personally before you returned to the United States.

Believe me, Colonel,
Very sincerely
(signed) John J. Pershing

Colonel George de Tarnowsky, M.C.
Medals awarded:

Distinguished Service Medal	U.S.A.
Victory Medal	U.S.A.
Medaille de Reconnaissance	France
Officier d'Academie	France
Medal of Olaf	Serbia
Star of Roumania	Roumania

Distinguished Service Medal
awarded
Colonel George de Tarnowsky, Medical Corps.

Letter from General Ireland.

Since my return from France I have had an opportunity to soberly reflect upon and make an analytical study of the problems that confronted the Medical Department Overseas, with a view to determining as far as practicable just what features in our organization were responsible for the good results achieved. The distinguished service of some officers of the Medical Department whose responsibilities and accomplishments overseas were of an outstanding character demand recognition. As the name of Colonel de Tarnowsky clearly falls within this category I urgently recommend that he be awarded the Distinguished Service Medal for his eminent services in directing the American Red Cross Military Hospital No 5, A.E.F. That the American Expeditionary Force was able to achieve success in the care of battle casualties evacuated to the Paris District during June and July 1918, was very largely due to the individual efforts and vision of this officer. In civil life, to which he has returned, he is a distinguished surgeon.

(signed) M.W. Ireland
Surgeon General U.S.A.

From General Mosely.

I hasten to congratulate you on the honor which has just come whereby you are to receive the D.S.M. I recall very well what a wonderful piece of work you did in organizing and commanding American Red Cross Military Hospital No 5 and I am glad to see Uncle Sam properly recognize your service.

(signed) E. VanHorn Mosely
Brigadier General

From General Ireland.

It gives me a great deal of pleasure to notify you that the Adjutant General has just communicated to me that the Distinguished Service Medal has been awarded to you for your splendid service during the war. I hasten to extend to you my very cordial, personal congratulations, and I am sure I am joined in these congratulations by all the officers conversant with your work in France.

(signed) M.W. Ireland
Surgeon General U.S.A.

Citation.
Distinguished Service Medal.

George de Tarnowsky, Colonel, Medical Corps, United States Army. For exceptionally meritorious and distinguished services in organizing and commanding American Red Cross Military Hospital No 5, American Expeditionary Forces. Notwithstanding the serious deficiency in personnel and material, Colonel de Tarnowsky, by his energy and efficient administration, received and treated great numbers of wounded evacuated to the Paris District during June and July 1918.

Official

The Adjutant General U.S.A.

Letter from War Department

Distinguished Service Medal – Surgeon General's Office Encouragement

August 13, 1921

IN REPLY REFER TO S.G.O.

WAR DEPARTMENT
OFFICE OF THE SURGEON GENERAL
WASHINGTON

August 13, 1921.

Lt. Col. George de Tarnowsky,
30 N. Michigan Avenue,
Chicago, Illinois.

My dear Col. De Tarnowsky: --

In view of the excellent services which you rendered during the World War, I have felt it my duty to bring your original citation for a Distinguished Service Medal again to the attention of the Adjutant General. The question of its approval now rests with the War Department Decorations Board. I trust that favorable consideration will be given to your case.

Sincerely yours,

M. W. Ireland,
Surgeon General

1 incl.

ALL COMMUNICATIONS SHOULD BE ADDRESSED TO "THE SURGEON GENERAL, U.S. ARMY, WASHINGTON, D.C."

Letter from War Department

Distinguished Service Medal – Approval to Award

May 24, 1922

GLD-eaw 1-214

WAR DEPARTMENT

THE ADJUTANT GENERAL'S OFFICE
WASHINGTON

IN REPLY

REFER TO A.G. 201 deTarnowsky, George
(5-19-22)WW

May 24, 1922.

Lieut. Col. George deTarnowsky
2719 Pine Grove Ave.,
Chicago, Ill.

Dear Sir:

Under the provisions of the act of Congress approved July 9, 1918, as amended by the act of Congress approved April 7, 1922, a distinguished-service medal was awarded you by the War Department on May 19, 1922.

The Quartermaster General of the Army has been directed to forward the medal to the Commanding General, Sixth Corps Area, 1819 West Pershing Road, Chicago, Illinois, who will cause the same to be presented to you with formal ceremony, including the attendance of troops.

It is requested that the Meritorious Services Citation Certificate previously awarded you by the Commanding General, American Expeditionary Forces, be returned to this office for cancellation, as the distinguished-service medal is a higher award than the certificate for Meritorious Services, and is awarded you on account of the same services for which the certificate was granted.

Very truly yours,

Acting The Adjutant General.

Letter from War Department

Distinguished Service Medal – Surgeon General's Office Congratulations

May 25, 1922

IN REPLY REFER TO S.G.O. 210.5-1

WAR DEPARTMENT
OFFICE OF THE SURGEON GENERAL
WASHINGTON

May 25, 1922.

Colonel George De Tarnowsky, M.R.C.,
30 North Michigan Avenue,
Chicago.

My dear Colonel Tarnowsky:

It gives me a great deal of pleasure to notify you that the Adjutant General has just communicated to me that the Distinguished Service Medal has been awarded to you for your splendid service during the War.

I hasten to extend to you my very cordial, personal congratulations, and I am sure I am joined in these congratulations by all the officers conversant with your work in France.

With cordial regards and best wishes,

Very sincerely yours,

M. W. Ireland
Surgeon General, U. S. Army.

ALL COMMUNICATIONS SHOULD BE ADDRESSED TO "THE SURGEON GENERAL, U.S. ARMY, WASHINGTON, D.C."

Record of Prior Service – Reserve Officers

Saturday, June 10, 1922

RECORD OF PRIOR SERVICE RESERVE OFFICERS REQUIRED BY ACT JUNE 10, 1922.

de Tarnowsky George			Colonel Med. Res
LAST NAME	FIRST NAME	MIDDLE INITIAL	RANK AND BRANCH OF SERVICE

To meet the requirements of the Act of Congress, approved June 10, 1922 (governing the pay of the Army, Navy, etc.,) I submit the following as a succinct statement of my COMMISSIONED SERVICE in the (1) Army (any component); (2) Navy; (3) Marine Corps; (4) Coast Guard; (5) Coast and Geodetic Survey; (6) Public Health Service; (7) Naval Militia; (8) National Naval Volunteers; (9) Naval Reserve Force; (10) Marine Corps Reserve; (11) National Guard or Organized Militia prior to July 1, 1916.

Commissioned 1st Lt Med. Reserve Corps U.S.A. Feb. 12th 1916
Appointed 1st Lieut Med. Reserve Corps U.S.A. March 17th 1917
Appointed Captain Med. Reserve Corps U.S.A. June 5th 1917
Appointed Major Med. Corps U.S.A. July 16th 1917
Appointed Lt Col.M.C., U.S.A. Nov 9th 1918
G.H.Q. Par 3.W.D. Cable 2147 Nov. 8th 1918 S.O. 313 Par 133
Appointed Colonel M.C., U.S.A. Feb 17th 1919.
Appointed Colonel M.C. Res, Jan 15th 1920; reappointed in 1925

SIGNATURE

RANK AND BRANCH OF SERVICE

(Address)

(THIS FORM TO BE MADE OUT AND FORWARDED TO THE ADJUTANT GENERAL OF THE ARMY WITHOUT DELAY.)

(For instructions see reverse side)

Instruction:

If service was rendered in No. 11, (National Guard, or in the Organized Militia prior to July 1, 1916,) a certificate of such service must be secured by the officer concerned from the Adjutant General of the proper State or Territory as soon as practicable and forwarded to The Adjutant General of the Army, Washington, D.C.

As the amount of Federal pay that an officer may be authorized to receive is dependent upon the facts sought to be elicited by fore-going inquiries, it is very important that this form be made out, signed and returned to The Adjutant General of the Army, Washington, D.C., as soon as possible, by each reserve officer accepting a call to active service, and also that there be no delay in procuring and forwarding the certificate from the Adjutant General of the State or Territory if commissioned in the National Guard, or in the Organized Militia prior to July 1, 1916.

Letter from Headquarters

July 6, 1935

HEADQUARTERS
Fort Sheridan, Illinois

DTM:lsh

July 6, 1935.

SUBJECT: Commendation

TO : Colonel George de Tarnowsky, Medical Reserve Corps.

1. I take great pleasure at this time, most probably the termination of your last tour of active duty with the Organized Reserves, in commending the excellent showing made by the 378th Medical Regiment, commanded by your-self, during its period of active duty training, June 23 to July 6, 1935, at Fort Sheridan, Illinois.

2. The enthusiasm shown by this unit is an indication that the energy and leadership which led to your assignment as Chief Surgeon, Ambulance des Allies, 14th French Army, as Commanding Officer, Hospital No. 5, as Commanding Officer, Balkan Medical Unit in 1919, and your designation to write Medical War Manual No. 7, "Surgery in the Zone of Advance", during the World War has not diminished with the passing years.

3. Your distinguished professional and military career have impelled me to recommend that, upon your retirement from service by reason of having reached the statutory age of sixty-four years, you be given a commission as a Brigadier General in the Auxiliary Reserve Corps.

DANA T. MERRILL,
Brigadier General, U.S.A.,
Commanding

Tribute

For Armistice Day

Friday, November 11, 1927

Later, when speaking of the war, he said, almost joyously, "I came through without a scratch!"

Written in scrawling cursive on linen parchment paper

For Armistice Day, November 11, 1927

He was not of unusual stature, but his Russian-blue eye spoke the strength, the courage, the endurance, the will-power of a Titan. These hero-making qualities, coupled with swift, unnerving judgment, top mastery of the service of surgery, and a towering sense of duty and devotion of his country and humankind, placed him in command of large and responsible positions on the firing-line of more than one front in the World War.

Day after day, week after week, month after month, through two age-long years, he never wavered in performing the task he had offered himself to undertake, though often exhausted and heart-sick to the breaking point. Through grueling hours, eight, then, twelve, nay, even eighteen without rest at critical times, his keen eye and mind viewed each suffering and bleeding man as he was brought before him, determined almost [instantly] what was needed to alleviate his distress and condition, performed the operation, ordered him to the proper place for recuperation, and turned to the next wounded soldier; always with the consciousness that any moment might use him, also torn as they were, or hurled into eternity.

But his high [pride] could not protect himself as it had battled to save his fellowmen. "He saved others; himself he could not save" from the attack of the dread infection of the trenches, and he at last succumbed, not knowing that he could even be restored to physical vigor. He was spared, however, and after many weeks was enabled to resume his work, and eventually to return to his home. Later, when speaking of the war, he said, almost joyously, "I came through without a scratch!" A man? Yea, verily. The great poet would have said of Colonel de Tarnowsky also: "...and the elements so mixed in him that nature might stand up and say to all the world, 'This was a man!'"

A Tribute to a Highly Respected Surgeon and Friend: Dr. George de Tarnowsky

Friday, May 15, 1942

"Few ever realized how much his knowledge of languages and the customs of the people, coupled with his winning personality, helped to smooth the strained relations of the Allied Armies."

-- Dr. P. J. Sarma

GEORGE de TARNOWSKY
1873-1942

On January 20, 1942, the surgical profession of America mourned the death of Dr. George de Tarnowsky which followed a brief illness. He was born in Nice, France, May 27, 1873, of a Russian-Polish father, Michel, and an American mother, Juliana Oakley, of New York. Both his parents were capable artists. His eldest brother, Michel, now living near Nice, is a distinguished sculptor. His youngest brother, Jacques, an engineer, died a few years ago in New Orleans. Doctor Tarnowsky survived his two sisters, Elizabeth (Madame Marcel Arnadon), a nurse in the first World War, and Marquerite (Madame Alexandre Durandy).

His elementary and high school education was received in Nice at the Anglo-American College. In 1891, a few months before his departure for the United States, he graduated from the Lycée de Nice of the University of France.

Shortly after his arrival in America he met the late A. B. Keyes, who persuaded him to take up the study of medicine. He entered Northwestern University Medical School and received his M.D. degree in 1900. In medical school he was a contemporary of Kanavel, Besley, Cubbins, Dyas, and other surgical leaders of our time. He served his internship at Mercy Hospital from 1900 to 1901, under Murphy, Andrews, Billings, and other men who made Chicago the "pantheon of modern medicine and surgery." Following his internship he entered private practice as an associate of the late E. F. Ingals, pioneer otolaryngologist of Chicago.

On September 3, 1902, he married Bertha Duffield Nixon, daughter of the late William Penn Nixon, onetime editor of the old Chicago Inter-Ocean and Collector of Customs, Port of Chicago. Mrs. Tarnowsky died in 1938.

The next year he spent in Europe doing postgraduate work in surgery. His knowledge of French, Italian, and German made it possible for him to get much valuable clinical experience during this visit. Among the European surgeons of that period, Bassini, Kocher, Pean, and VanHacker left great impressions on his mind. He revisited European clinics during 1912 and 1913.

In 1907 "four Georges" (George Green, George Busse, George Baxter, and George de Tarnowsky), organized Ravenswood Hospital. Doctor Tarnowsky was a senior attending surgeon of this institution until the time of his death. One of his proudest achievements was his contribution to the high scientific and professional position which Ravenswood holds today. His loyalty, devotion, and untiring efforts to advance the practice of surgery at Ravenswood Hospital were an inspiration to the men who were associated with him.

Doctor Tarnowsky was an instructor in gynecology at the Chicago Poli-clinic from 1905 to 1912. He joined the faculty of Loyola University School of Medicine in 1913, and held the rank of clinical professor until 1926.

For almost twelve years he was attending surgeon at Cook County Hospital. His clinic there was a Mecca for medical men. In the operating room one was inspired by his simplicity of technic, directness of approach, gentleness, meticulous care in handling tissues, neatness, and rapidity. Surgical operation was a sacrament, and he realized the serious responsibility of life entrusted to him.

He held the position of professor of surgery at the University of Illinois College of Medicine from 1926 to the time of his death. As a teacher, there was none superior on the University of Illinois faculty. His mastery of languages, profundity of surgical knowledge, ability to separate the chaff from the grain of Himalayanic surgical literature, and his ability to present a most difficult subject in a simple, interesting, and retentive form, made him one of the most popular teachers of our time. He was one of the last amphitheater clinicians. His clinical lectures were attended by large audiences. Although many of his courses were elective and the "roll call" system was never adhered to, an empty seat was hard to find in the amphitheater during his lecture hour to the students.

Medical students, interns, residents, and practitioners had no better friend than Doctor Tarnowsky. He was always eager to give a helping hand to young men in the medical profession. The confidence of his medical colleagues in his surgical judgment and ability demanded much of his time, which he gave cheerfully and freely.

His contributions to the literature were many. Over a hundred articles are to his credit in various medical journals in America and Europe. His "Surgery of the Zone of the Advance" (Medical War Manual No. 7) was a surgical bible to the American medical officers during the World War. "Emergency Surgery," published in 1926, received an enthusiastic international reception. The first article written by him during his internship at Mercy Hospital was on "General Anesthesia," and for it he was highly commended by Murphy and Andrews. His observation and publication of "Ruptured Bladder" was considered the most masterly presentation on that subject. He also served as abstract editor of the

International Abstract of Surgery from 1921 to 1922. In the "Medical and Surgical History of the World War," he contributed a chapter on "Surgery of the Front."

He was a fellow of the American Medical Association, Institute of Medicine of Chicago, American College of Surgeons, and a member of the Chicago Gynecological Society and Chicago Surgical Society. He served as president of the two latter organizations from 1937 to 1938 and from 1938 to 1939 respectively. He was a founder member of the American Board of Surgery and a member of the Illinois State and the Chicago Medical societies, and of Nu Sigma Nu and Sigma Xi. He was also a member of the University Club of Chicago.

The army career of Doctor Tarnowsky was the most colorful chapter in his life. Soon after our declaration of war in 1917, he was in training at Fort Benjamin Harrison. While there he was chosen by the editorial committee, Council of National Defense, to prepare a book on war surgery. The various chapters of this book were outlined by him in the United States from data at hand. He then proceeded to France to gather firsthand information from his English and French colleagues and from the battle front. This was published in book form, "Medical War Manual No. 7, Military Surgery of the Zone of Advance," and was commended highly by military authorities. He was later chief surgeon of the Ambulance des Allies, and organized and commanded American Red Cross Military Hospital No. 5.

Doctor Tarnowsky was promoted to the grade of colonel on February 17, 1919. Numerous decorations were awarded him by various governments for his meritorious service, but none was cherished more than the Distinguished Service Medal awarded him by the Government of the United States. The French Government made him a "Chevalier de la Legion d'Honneur" and decorated him with the "Medaille de Reconnaissance" and "Officier d'Academie." Few ever realized how much his knowledge of languages and the customs of the people, coupled with his winning personality, helped to smooth the strained relations of the Allied Armies.

In January, 1919, after his work with the Red Cross Hospital No. 5 was finished, Doctor Tarnowsky was sent to the Balkan States by the American Red Cross for reconstruction work. He did not return to the United States until September, 1919, and was discharged from the Army in that month. After the war he organized the 378th Medical Regiment, and served as commanding officer until his retirement to inactive status in May, 1937.

During his last visit to Europe in 1939, Doctor Tarnowsky became convinced of the imminent danger of Nazi attack on peaceful democratic nations, and after returning to the United States he stressed the importance of thorough preparedness. He offered his services to the Surgeon General of the Army as soon as the National Emergency was declared and, there is no doubt, had he lived he would have been called to active service.

Doctor Tarnowsky was a man of simple tastes and high ideals. He was fond of music and the out-of-doors. Mountain climbing and tennis were his favorite sports. He was never more happy than when among the members of his family and friends. He is survived by two daughters, Olga (Mrs. William Hudson) and Mary, two sons, Alexis and Nixon, and five grand-children.

Two of his most distinguished characteristics were honesty and usefulness. Truth was a religion to him. His philosophy of life could be best expressed by a quotation from Keat's "Ode on a Grecian Urn":

"Beauty is truth, truth beauty," -- that is all
Ye know on earth, and all ye need to know.

P. J. Sarma

Reprinted from the Proceedings of the Institute of Medicine of Chicago
Vol. 14, No. 5, May 15, 1942

Miscellaneous Images of Dr. George de Tarnowsky

Red Cross Hospital #5

I have just been informed that there are only three vacant beds left! More cots were procured! July 1918

Convalescents were out to see if any of their pals have arrived. [My sister] Bessie and I both in the picture. July 1918

One of my many duties, Saturday morning inspection of my staff and Hospital Corps men. July 1918

Severely wounded on their stretchers, waiting to be assigned. [My sister] Bessie in white. July 1918

About the Author

Alongside writing *Military Surgery of the Zone of the Advance* (published June 1918) as designated by the Editorial Committee, Council of National Defense, eminent surgeon Dr. George de Tarnowsky from Chicago expanded his ethnographic and phenomenological study through riveting storytelling, perceptibly illustrating the wartime and postwar landscape of a melting pot imbued with noble diplomatic relations of Allied counterparts—and even adversaries. With letters, orders, and other documentation accompanying this compilation of essays, the French-born American immigrant son of a Russian-Polish father and an American mother depicted his World War I and Post-Armistice missions with multilinguistic proficiency, tapping away on a typewriter or hastily scrawling in his journal in a myriad of settings—from the dimly lit hours of the early morning following grueling days of surgery to the rhythmic rocking motions experienced on both rail and water.

Dr. de Tarnowsky returned home to resume his medical career, surgical research, and instruction of aspiring surgeons. With accolades from the royalty of France and Romania and recommenced efforts from the Surgeon General, Dr. de Tarnowsky was honored with the Distinguished Service Medal for extraordinary wartime services.

Made in United States
Troutdale, OR
11/03/2026

56845379R00215